D1605790

FULLER'S EARTH

A Day with Bucky and the Kids

By Richard J. Brenneman

ST. MARTIN'S PRESS
NEW YORK

Drawings and photos by the author.

FULLER'S EARTH: A Day with Bucky and the Kids.
Chapters 1, 2, 5, 6 copyright © 1984 by Richard J. Brenneman. Chapters 3, 4 copyright © 1984 by Richard J. Brennenman and R. Buckminster Fuller. All rights reserved. Printed in the United States of America. No part of this book may be used or reproduced in any manner whatsoever without written permission except in the case of brief quotations embodied in critical articles or reviews. For information, address St. Martin's Press, 175 Fifth Avenue, New York, N.Y. 10010.

"Memories of Bucky" by Norman Cousins. © 1983 Saturday Review Magazine Company, Inc. Reprinted by permission.

Library of Congress Cataloging in Publication Data

Fuller, R. Buckminster (Richard Buckminster), 1895–1983
Fuller's earth.

Bibliography: p.
Includes index.
1. Technology—Social aspects. 2. Civilization.
I. Brenneman, Richard J. II. Title.
T14.5.F934 1984 620 83-19219
ISBN 0-312-30981-3

First Edition

10 9 8 7 6 5 4 3 2 1

With humility and gratitude I dedicate this book:

—To Richard Buckminster Fuller, for his extraordinary dedication to humanity and for his enthusiastic desire to share his thoughts through this book,

—To Laura Shelton Brenneman, for the love and gentle encouragement without which this work would never have been published, and

—To my parents, for everything.

I would also like to thank Benjamin Mack, Rachel Myrow, and Jonathan Nesmith for their invaluable role in this project, and for being the bright lights they really are. Special thanks are due also to James H. Craft, a remarkable teacher who imparted to at least one student his love of language.

—Richard J. Brenneman
Davis, California
July 5, 1983

Contents

Foreword: *MEMORIES OF BUCKY*

By Norman Cousins

Once when we were in an American delegation to Moscow for the purpose of exploring issues between the two countries, Bucky Fuller gave a talk on the human future that no one present soon forgot.

One of the Russians had suggested that as a relief from our regular conference sessions, we might stage a debate on what the world would be like in 2000 A.D. Their debater would be Eugene Fyodorov, the famous meteorologist and futurist. Inevitably, for our forensic gladiator we picked Buckminster Fuller, architect, inventor, cosmic chronicler, philosopher, and poet.

I groaned when I learned the ground rules set up by the Russians. Each debater would have fifteen minutes, *precisely*. I had never heard Bucky speak publicly for less than two hours. When the rigid time limit was explained to Bucky, he shrugged, giving us the impression that fifteen-minute talks were a matter of casual routine.

Professor Fyodorov spoke first. Systematically and methodically, he presented a checklist of all the factors that he believed would bear on the world's economy in 2000 A.D. He extrapolated figures with respect to world population, world food supply, world supply of vital resources, etc. I looked around at the Russians as their champion spoke. They were obviously pleased with the coldly scientific and comprehensive nature of the presentation.

After fifteen minutes, give or take ten seconds, Professor Fyodorov completed his talk and sat down. Substantial applause from all present.

Bucky started to speak. Within three minutes, he cast a spell over the entire group. The Russians sat forward in their seats. The world's greatest resources, he said, were to be found in human intelligence, ingenuity, and imagination. He identified the principal problems of the riders on Spaceship Earth and gave the reasons for his belief that these problems were well within human capacity to solve. His earnestness, enthusiasm, creativeness, and knowledge were beautifully blended.

Bucky sailed through the fifteen-minute barrier with the ease and confidence of Roger Bannister going through the four-minute mile. As chairman of the evening session, I started to rise to inform Bucky his time had expired. I felt a restraining hand on my arm. "Please let Mr. Fuller continue," Professor Fyodorov said. "He is magnificent, absolutely magnificent. You must not stop him."

I settled back in my seat. Bucky continued for almost an hour. The Russians were mesmerized. In the midst of the applause following his talk, Professor Fyodorov whispered in my ear: "It was no contest. Mr. Fuller is the winner. Never in my life have I heard anything so wonderful. I am sorry he stopped so soon. Tell me, what did he say?"

The professor was not being sarcastic. Audiences all over the world have had the same experience. They may not have known or understood quite what Bucky was saying, but they felt better for his having said it. He gave people pride in belonging to the human species. He gave them confidence in their innate abilities to overcome the most complex problems. He made them feel at home in the cosmos.

Very early in life he discovered the secret of perpetual curiosity, and he spent the rest of his life trying to give the secret away. Of all his attributes, none was more compelling than his ability to transmit to others his kinship with the universe. His uniqueness as a teacher in this respect was that he saw poetry in everything. He viewed physics, astronomy, chemistry, and other sciences as much through the creative imagination as through equations and formulas. In so doing, he refuted C. P. Snow's notion of the gulf between the "two cultures." He regarded science and the fine arts as extensions of each other, as manifestations of an integrated reality.

The affection of students for Bucky Fuller offered the strongest possible evidence that young people are responsive to the values that give affirmative energy to a society. What they understood through him was that the main end of science is not to answer questions but to generate new ones: not to relieve curiosity but to enlarge it and ignite it. I have known very few people who, after meeting Bucky, did not forever feel a sublime wonder when looking at a starlit sky.

If we read Bucky Fuller solely for information we will obtain information, but we will be cheating ourselves. We should read him for the increased respect he gives us for human potential, and for the lesson that there are no boundaries to the human mind, which he celebrates above all else. The great poets have attempted to describe the human mind and spirit, but I doubt that any of them have done so more provocatively than Bucky. The reason perhaps is that Bucky was not only inspired and nourished by the weightless and all-embracing entity called the human mind, but he had a way of opening our minds to the phenomena within them. In this way, he introduced us to ourselves.

PART ONE
Note to the Reader

"I think it would be wonderful if someone could put you together with some children and then record the results," I ventured. "The theme should be 'What information do today's children need to help them through the coming years?' "

"Go ahead," Bucky responded. "I'm at your disposal."

And thus a book was born.

Richard Buckminster Fuller was one of the most remarkable individuals who has ever lived. His wide-ranging mind, hauntingly free of the sense of limits that afflicts most of us, conceived poetry, architecture, automobiles, boats, furniture—and some of the most provocative and original mathematical, geometrical, and philosophical speculations ever written.

There was something marvelous about Bucky (you couldn't think of him by any other name once you had met this elf disguised as an octogenarian). His most infectious quality was an unfettered sense of the immense potential of all humans. It was almost impossible to be around him without being caught up by it.

He greeted all comers openly, with an evident love. You were his peer, whether you were a physicist, journalist, mail carrier, or child. The equality he offered was difficult to ignore.

As you listened to Bucky, you rediscovered the art of the storyteller, of the painter of word pictures who can excite the imagination into visions of new worlds filled with joyous new possibilities. You would be fascinated by his conceptions of humanity, universe, and the Greater Intellectual Integrity that to him lay at the root of all experience. If you let yourself, you would share his dreams and quietly exult in the serenity that flowed from his certainty in the ultimate *goodness* of things—although he would have also told you he eschewed the use of words like "good" and "bad." You may have disagreed with particulars, but you sensed something *right* in his sense of the grandeur of humanity, and of our place in the cosmos.

And you would have delighted at the incredible range of his own accomplishments, and at your growing understanding of the marvelous coherence of the thought that had given rise to them.

You would have recognized that at the heart of his vision was the conviction that whatever *he* had accomplished was within the reach of *any* human—an absolute certainty that man exists as the expression of the divine creativity, endowed with the potential of dominion over every aspect of experience. And as he gently smiled and touched your hand, calling you "dear one" with an utterly unaffected and ever-so-gentle warmth, you were drawn more deeply into his magical world. You were his equal, a fellow player in some cosmic game, sharing precious moments together.

It was only later, after you had left, that the full impact of that equality would have sunk home. Good heavens, the thought would have come. If he's right—if he's only an "average human being" who has devoted his life to proving what an "average human being" can do—then what have *I* been doing with *my* life?

Such gentle but profound shocks may prove the source of Buckminster Fuller's most enduring impact. This tender-hearted yet fiercely tenacious man encountered thousands of people in his life, and many found themselves touched by that intense yet childlike conviction of what it means to be fully human.

Here was a visionary, an old-fashioned mystic utterly certain of the here-and-now possibility of ecstatic communion between man and God. There was much of the New England Transcendentalist in him—his great-aunt Margaret Fuller had been one of the pillars of that uniquely American movement.

The intensity of his vision flowed from his own mental powers, honed by a fierce sense of discipline. He slept but little, devoting his waking hours to prodigious mental labors. Here was a man who truly lived. His life was an endless vortex of activity, and even in his last years he maintained a pace that would have taxed the energies of anyone half his age. There were meetings, planning sessions, and consultations; talks at college commencements, and "est" gatherings; and, at the end, the Integrity Days.

His most famous invention, the geodesic dome, is everywhere, literally, from the poles to the Equator, enclosing more space with less material than any structure previously devised by humanity. The impact of his thought is equally dispersed in the lives of those he has touched. And Bucky Fuller was a man who devoted his life to serving others.

This book arose out of an interview with Bucky conducted while I was a reporter for the *Evening Outlook,* a newspaper in Santa Monica, California, just a few miles from the house in Pacific Palisades that served as home base for Bucky and his wife, Anne, during their last four years. Bucky had picked the coast so he could be near his daughter, Allegra Snyder, and grandson, Jaime. (A granddaughter, Alexandra Snyder, lives on the East Coast.) His office, however, was in Philadelphia, and Bucky himself could be found almost anywhere. If you really wanted to keep track of this peripatetic globe hopper, you could have subscribed to his printed scheduling service.

If there is any single quality in the characters of those "great" men and women I have been privileged to meet, I would have to say it is a childlike openness to some aspect of the universe. That quality was present in Buckminster Fuller to a degree I had never seen before. There was a directness and lack of pretense that left me almost in awe. It was not that he was devoid of ego—he had a very healthy sense of self—rather, his ego was subordinated to a transcendent cause, which he described simply as "making the world work." It was this cause that fired the imagination of Fuller and those around him.

And it was his childlike clarity of purpose that led me to

propose the encounters that have culminated in this book.

As a reporter, I had discovered that I was serving as a translator for Fuller. In his communications with audiences numbering from one to thousands, Bucky constantly strove to draw the maximum exertion from his listeners; he would communicate at their highest ability to follow. As one who had read most of his books and followed his career, my ability to tolerate his complex and peculiar linguistic usages was high. Yet to reduce the concepts to the level demanded by journalistic constraints demanded that I serve primarily as a translator. There are always dangers in translations, I realized, for the translation is only as good as the translator's own grasp both of the content of the original and of the language into which the original is translated. Some degree of error inevitably creeps in. By using children, I would be forcing Bucky to translate himself; and there could be no better guarantee of accuracy.

Bringing Bucky together with children, I reasoned, would accomplish several things. First, it would prevent him from using those complex and idiosyncratic terms that can throw off all but the most dedicated reader or listener; second, it would provide three children with an opportunity to spend time with a remarkable being; finally, the encounter could provide the basis for a book that would serve as the most gentle introduction to Fuller's universe.

I selected three youths for this experiment. My only criterion was that they possess inquiring minds and an ability to grasp new concepts.

Rachel Myrow, ten at the time of the first session, is the youngest. She is the daughter of Fred and Illana Myrow, respectively a Los Angeles composer/producer and an Israeli-born actress. Benjamin Mack and Jonathan Nesmith were both twelve. Ben is the son of Lucy Mack, a Los Angeles–based film producer, and Art Mack, a teacher. Jonathan is the son of Phyllis Nesmith, a Los Angeles political consultant, and Michael Nesmith, a musician and media expert of Carmel, California.

I gave the children one charge: They were to consider themselves representatives of all children everywhere, and they were to ask questions they thought would reflect the concerns of all children. I asked Bucky to hold one thought during the sessions: What should children know to help humanity through the planetary crises now becoming so apparent.

Bucky met with the children for three sessions over a period of two years. The results were tape-recorded and captured by still photography. This book is the result. The text has been edited, sometimes extensively, to compensate for those jumps in understanding that can come when people are communicating face to face. In some instances, basic concepts have been elaborated. Drawings have been added, some based on sketches executed by Bucky as he spoke to augment points in the text, others, as needed to impart clarity. Any additions in drawings or text have been made in the spirit of the actual event, as the children—who have read the manuscript—have attested.

Before their first encounter with Fuller, I prepared the children with two "briefing sessions," at which I presented my own understanding of Fuller and his world, gleaned from my own long-standing interest and extensive reading. I played tapes of my own interviews with Bucky, showed models of his geometric concepts, explained the Dymaxion map, and let them look through some of Fuller's many books. From this foundation, the children prepared their original questions.

After the first session, I also asked the children to record their own impressions of the event.

I know the encounters were meaningful to the children, and Bucky referred to them frequently during his last speaking engagements. He told me he had never been asked better questions.

This book is the fruit of three encounters between four diverse yet remarkably similar individuals. It is also an introduction to a strikingly different way of looking at the world. Many of the concepts may seem startling. Don't be put off if some idea seems at first difficult to grasp. Fuller's ideas prove to be both consistent, coherent, and understandable, as the three children discovered. The index at the back of the book contains a glossary of terms used by Fuller, and there is a bibliography of his works should you choose to explore his universe more deeply.

This book is meant to be enjoyed. It was enjoyable to produce, enjoyable to write. I hope it's equally enjoyable to read.

PART TWO

Fuller's Faith: An Introduction to Bucky Fuller

It was 1927, the year a bashful young pilot named Charles Augustus Lindbergh fired the world's imagination by making his solo flight across the Atlantic Ocean. It was a very different world from the one we know today. There was no television; and until that very year, all the movies ever filmed had been silent, save for the impromptu musical accompaniment played live on organs and pianos in theatres where the films were shown.

America seemed prosperous and at peace. There was an unprecedented economic boom, and just over the horizon shone an era when *everyone* would be rich—a dream fueled by an ever-accelerating bull market. The music heard on the radio (still a relatively new invention) was happy. Two of the year's biggest hits were "Blue Skies" and "Let a Smile Be Your Umbrella." It was also the year the fifteen-millionth "Tin Lizzie" rolled off Henry Ford's assembly lines.

The twin horrors of the Great Depression and World War II lay ahead, in a future undreamed of save by a few savants labeled doomsayers and gloom spreaders.

But for the young man standing on the Chicago shoreline, life looked hopelessly bleak. He was a small bulldog of a man, short and broadly built with a large head. He walked with an erect, almost military bearing that struck some as Napoleonesque. His jaw was prominent, commanding. He peered out at the world through bottle-thick lenses that magnified his already owlishly wide hazel eyes. He had a broad patrician mouth that could break into a dazzling, infectious smile when he was happy.

But there was no joy on this bleak night. Had any passerby

noticed, he or she would have seen a face racked with anguish.

Bucky Fuller had decided to kill himself.

He was going to swim out into the icy dark waters until he could go no farther. Then he would surrender himself to the waves. It would be easy, he reflected. He had always loved the water, and there was something oddly reassuring in the thought of surrendering to its final embrace. Life had become unbearable for this gentle, loving man.

If there was a single root cause for his misery, it lay in his inability to see things the way others did. Oh, he had tried—and hard. But somehow, every time he accommodated himself to the insecurities and fears of others, he found himself once again confronting the specters of failure and financial ruin.

It had started in school, with his incessant questioning of teachers. *Why* and *how* mattered deeply to him, and he could not content himself with the glib explanations provided by men and women who had obviously never asked themselves the same basic questions the young Fuller was always asking himself. The teachers often seemed upset and provoked by his inquiring mind, seeing in this small bespectacled child a threat to their own authority. He was early branded a troublemaker. But the teachers failed to recognize that all he wanted was to *understand.*

Somehow he managed to adapt and make his way through primary and secondary school.

Richard Buckminster Fuller came from an old New England family. For five generations, Fuller men had graduated from Harvard. When it came time for Bucky to go to college, he followed in the family tradition.

It may have been that being branded a rebel turned him into one. For whatever reason, the young Fuller found himself engaged in all sorts of benign mischief (including one remarkable evening when he spent an entire semester's allowance on a champagne dinner for a troupe of Broadway chorines). Such antics were poorly thought of in the tradition-bound halls of Harvard, and Fuller was expelled.

After his ouster from academia, Bucky found a job as a machinist, thanks to the intervention of a kindly relation. He loved the work, especially the almost alchemically magical alloys—those strange combinations of metals that turned out far stronger than the combined strengths of their individual components. They hinted to the young Fuller of a world unseen by the senses where forces only vaguely discerned might hold untold promise (or menace) for the future.

He managed to win his way back into the university's good graces for a second time—only to find himself booted out again. That was to mark Fuller's final assault on the corridors of formal education. It wasn't that he didn't want to learn. It was just that the things he was being taught didn't make that much sense. And, besides, there were all those lovely young chorines, all that exciting *life.*

His happiest years came during World War I, the "Great War" or the "War to End All Wars." During the war, Bucky served in the Navy, first commanding a family-owned boat that had been volunteered as a vessel for seeking out German submarines along the New England coastline.

Bucky's inventiveness surfaced during his military years. One assignment involved the rescue of pilots of the day's primitive aircraft. When the biplanes crashed into the water, the craft often flipped over, drowning the trapped pilots. Fuller promptly designed a mechanical arm for righting the planes, saving the lives of countless airmen. Fuller's brilliance drew the attention of superiors, and he was selected to attend a three-month crash course at the U.S. Naval Academy at Annapolis. He emerged an officer.

It was in the Navy that Fuller discovered a way of thinking about the world that was to profoundly influence the course of his life. He became viscerally aware that the world was a single entity, not a collection of nations separated from each other by imaginary lines. This may sound strange today, to us who have seen the world as it appears from the surface of a moon a quarter-million miles distant. But in those days, humanity was confined to the surface of the earth. Airplanes were still novelties, held close to the ground by the limitations of the day's technology. Maps were generally multihued pastel affairs where each nation bore its own color, and differences stood out far more than commonalities.

But ships sailed the seas freely, traveling continent to continent over the mantle of water that was the one-world ocean, despite the arbitrary labels assigned by man to its various aspects. From the cabin of a ship, the world is a single vast expanse of water over which move vessels filled with food, clothing, metals, fuels, machines, and a host of other manufactured goods and raw materials.

This new perception of the earth was to have a profound effect on Fuller's thought. He came to realize that our planet itself was like a ship, containing vast but nonetheless limited resources, like the food and fuel carried aboard an ocean-going ship. Like the seafaring vessel, the earth also carries its complement of passengers—each living thing that dwells in or on its soil, water, and atmosphere. Seeing our world as "Spaceship Earth"—a term Bucky later coined to describe this way of looking at things—convinced him that each of us must take more care to think about how our actions affect other beings, human and non-human. This was decades before the popularization of the word "ecology."

During the war, Bucky had married the one and only love of his life, Anne Hewlett. Theirs was a formal naval wedding, complete with a recessional march under the crossed sabers of fellow officers clad in their resplendent dress whites.

Toward the close of the war, Bucky was involved in a momentous experiment. His tiny ship ferried the inventor of the voice-carrying radio out into a calm bay where history's first conversation between aircraft and ground—or ship, to be more accurate—took place. Fuller was present at the birth of an era. When the war ended, the newly married officer was assigned to serve on the communications staff of President Woodrow Wilson as he sailed to the international peace conference in Paris. Bucky made two trips aboard the presidential ship, and during the second was present in the communications room for the first radio broadcast to carry a human voice across an ocean. For Fuller, it was an exciting moment, presaging an era when all the passengers of Spaceship Earth would be linked in a network of virtually instantaneous communication.

Not long after Fuller's return from his second European voyage, Anne gave birth to a daughter, whom they named Alexandra. The child soon became very ill. The condition was diagnosed as spinal meningitis. The child later contracted polio. Physicians ordered round-the-clock nursing care, holding out little hope for the infant's life. Bucky was heartbroken.

Then came a second blow. The Navy told the young officer he must prepare to ship out for the Far East. Bucky tried to win reassignment to an East Coast port, because Anne needed to be near her family and the young father did not want to leave his wife and daughter during the height of their crisis. But the service proved unrelenting, and with deep reluctance Bucky resigned his commission.

His obvious intelligence and military administrative experience won him a position with the Armour meat-packing company as chief of their importing section. An offer from another firm followed quickly, with the promise of higher pay. The promise of more money to meet the family's mounting medical bills led Fuller to jump at the offer. But disaster struck when his new employer went bankrupt soon after Fuller came aboard, leaving the young family worse off than before.

And then Alexandra died.

Bucky was devastated, for two reasons. First, because he had come to dearly love the little child; second, because he had been away at a football game at the time she died, on her fourth birthday. His absence laid a burden of guilt on his shoulders so profound that his voice still choked with emotion as he recounted those events six decades later.

Somehow he managed to continue. He found himself a new position, this time as the president of his own company. Anne's father had invented a new system of building with bricks, involving a radically new technology of brickmaking, coupled with a system Bucky helped perfect for laying the improved bricks that would save both time and money. A corporation was formed, with Bucky at the helm. The Fullers moved to Chicago to be near the brickmaking factory, then under construction. For a time, the future seemed bright; but in 1927 Anne's father died, and internal corporate problems, coupled with government regulations and resistance from unions and established builders, blocked successful adoption of the new building system. Anne gave birth to their second daughter, Allegra, just as the company collapsed. Bucky was again without a job.

For the thirty-two-year-old Fuller, life held nothing but despair. He was hopeless, a failure. The only time he had ever felt satisfied with his work was in the Navy, which personal disaster had forced him to leave. He had struggled in the following years, but to no avail. He had nothing to offer the world. And there was the numbing, gnawing guilt he felt for his absence at the time of Alexandra's death.

As his thoughts gathered momentum on their downward spiral, Fuller realized that his own mother and Anne's mother would be able to care for his wife and child should anything happen to him. And his own record of apparent failures convinced him that he could bring nothing but more suffering and pain to those who loved him. He was left with the conviction that his best course of action would be to "get himself out of the way." To commit suicide.

It was this confusion and pain that brought him to the shore of Lake Michigan that night in 1927. But as he stared across the wind-tossed surface of the waters, something happened, something that would eventually touch millions of lives and provide other desperate souls with a sense of new possibilities, of a world free from want and deprivation.

And it all started with a single thought: "This is the last time you'll ever have to use your mind, so you'd better use it. You'd better do your own thinking—see what you *really* think."

It was an infinitely precious moment, a timeless epiphany in which a single current of thought poured in with overwhelming lucidity, sweeping every other thought and feeling before it with an irresistible rush of clarity. In that soul-transfixing moment, Bucky suddenly realized that much of his pain had come from his sincere efforts to *believe* what others had told him to believe, regardless of his own contrary feelings and intuitions.

"I'd been brought up by an older generation absolutely certain that the mental processes of the young were unreliable. I'd been brought up continually hearing the expression 'Never mind what *you* think, this is what *we're* teaching you.' I'd been taught to play a game.

"And I saw that the people who were telling me to do that—to do what they were telling me—apparently loved me. I knew my mother did, and I felt that most of the others did too. So I had decided that I would do my best *not* to do my own thinking, to play by the rules as children learn to do. Everyone had been telling me to 'get over my sensitivity,' to 'realize that life is a hard battle.'

"Yet I had also observed that many, many times what I had been thinking turned out to be true, and what society was telling me to believe turned out not to be true."

So there it was. The thought. See what *you* really think. Think for yourself.

Another thought came.

Much of the pain had also arisen from selfish acts, when he had placed his own "needs" over those of others, when he had approached life with the attitude of "What's in it for me?" From some deep, intuitive wellspring came the realization that selfishness was the source of anguish. Therefore, why not change your thinking, he reasoned with himself.

"What if I look at life differently? Suppose I commit my life from now on never to *me* anymore, but instead use my life and experience only for others?

"I thought, Who are we? What is life?

"And then I realized that each of us is an incredible inventory of experiences, and that I might be able to use some of my experiences so that others would not be hurt the same way I was hurt, so that I could help others from coming to the same pains I had come to."

A vast reformation was taking place in the young man's consciousness. Time seemed to have ceased, and he was aware only of the momentum of his own thoughts, coursing deeper and deeper toward some still-unseen but life-renewing center.

Another thought came. If pain had come from acting on erroneous beliefs others had instilled, what was he *now* to believe? Or was he to *believe* anything at all? And if not to *believe,* then what?

Bucky resolved to take his lead from the sciences, where a rigorous methodology was prescribed for arriving at truth. He would implement the scientific method in his own life. He would function no more on unsubstantiated beliefs. A "belief" was unscientific, a hand-me-down superstition which could only cloud and obscure the face of reality. In those areas of experience where he had no conclusive proofs, he would be an honest, meticulous scientist, evaluating the evidence of experience and formulating hypotheses which he would then set out to prove in the crucible of his life. Nothing would go unquestioned, and where he had no answers of his own he would tolerate ambiguity until scientifically perceived experience wrought an answer.

Now his thought was approaching the one question he knew was central.

"One of those things you've been taught to believe is about God. If you're going to think for yourself, then what experimental evidence do you have to assume that a Greater Intellect than that of humans is operating in Universe?"

With his scientific training, Bucky had come to see that the universe of human experience is governed by *laws*—laws that are weightless and invisible in themselves, yet operating everywhere we look, in every experience, with complete and absolute power. These are the laws that hold the colossal suns and planets in exquisite balance, the laws that govern the intricate, invisible realm of the atom.

"And I saw that these laws could only be expressed mathematically. So I said to myself, 'Mathematics is intellectual, mental, and these laws are expressions of intellectual principles that predated humanity's discovery of them.' And I saw that it was quite apparent that there is a Greater Intellect than that of humans operating in Universe," the Intellect that expresses itself through the laws governing all experience.

And then there was the ultimate realization, that of the ever-presence of love. Here, in love, Bucky saw, was further evidence of a power operating beyond the reaches of human understanding. "I thought a great deal about love at that moment, about how incredible the phenomenon we call *love* really is. Stones don't love stones, yet humans take love for granted. And I thought about children, and how as children we all feel it so much. It's absolutely weightless, and it's an incredible phenomenon.

"I knew then that there's something operating here that's way beyond our understanding."

(Bucky's eyes lost their focus as the memory of that transcendent moment overwhelmed him. He brought his hands together, as if in prayer. His eyelids closed, and the flow of words ceased. The power of that decades-old vision retained its hold on his imagination.)

Suddenly, on the shore of Lake Michigan, Bucky realized that nothing had been lost. There was nothing in his life, no experience at all, that could not be transformed into something that would benefit others. His years as a machinist had taught him about a realm of invisible yet incredibly powerful forces ready to serve the needs of humanity. His years in the Navy had taught him to see the world as as unfragmented whole, as Spaceship Earth with all humanity as passengers sharing a common destiny. His apparent failures in the business world had taught him the limitations of thought governing both the commercial and governmental spheres.

At that moment, Buckminster Fuller's life ceased to be a collection of good and bad memories; in its place was a wealth of experiences, some pleasurable, some painful, and each usable to the benefit of his fellow passengers on Spaceship Earth.

"I said to myself that I have to realize that I don't know why I'm here, and that it's not really for me to say that this particular inventory of experiences should be removed from the availability of others. I realized that we are born to help others, and that each of us is related to all humanity. I saw that if I really looked at myself as some part of the grand design, I might represent in my experience some links that might be useful for humanity, and that might be lost if I were to throw myself away."

Bucky's unique "inventory of experiences" had equipped him as a thinker and designer. He knew about tools, metals, machines, technology. In the Navy, he had seen this knowledge turned toward destructive ends, employed to annihilate fellow humans and their life-supporting (as well as war-making) technology. Bucky resolved at that moment to turn the technology of *weaponry* toward a new aim—*livingry,* the solving of humanity's life problems. He would help others by turning his skills to the development of solutions to such basic human problems as housing and transportation. And he would not confine his interests and explorations to a single, narrow speciality. Life and Universe* itself would be his subject. He would become a *comprehensivist,* always endeavoring to see the broadest possible picture. There would be no area of experience outside his interest.

* To Fuller it was always "Universe," never "the universe," just as to a monotheist it is "God," not "the god." To Fuller, Universe was one and indivisible, the manifestation of the Greater Intellectual Integrity.

There was one last element needed to complete his transformation, one final thought.

Looking back at his own experience and at the collective portrait of human experience called history, Bucky realized that most of humanity's worst actions were motivated by fear, specifically by the belief that the world contains limited resources, that "there's not enough to go around." Deep within virtually all the earth's established systems of belief lurked the "law of the jungle." A man or woman had no inherent right to exist, to live. This was the basic credo of human fear. You had to *earn* a living, either as the producer of material goods or as the supplier of services to a producer—(be it as employee or spouse). Some would win; others were destined to failure—and death. This same sense of "lethal inadequacy of life support," a phrase Fuller later coined, was at the core of the prevailing strains of political, religious, and social thought. In the modern world it reared its head as the "class struggle" of Marxism, the "survival of the fittest" of Darwinism (social and biological), and in myriad other forms.

Yet Bucky saw that humanity had also harnessed principles enabling the performance of ever-greater works with ever-smaller amounts of matter and energy. This process of doing more with less he labeled "ephemeralization." Man had progressed from track-bound trains to trackless trucks and automobiles and on to the airplane. The speed of travel had greatly increased, while the amount of matter and energy needed to perform the act of travel had diminished in almost equal measure. And so, too, had humanity moved from ground-dependent to wire-dependent communications. Fuller himself had witnessed the birth of wireless voice communications from ground to air and from continent to continent. Now, in 1927, it was becoming possible to communicate with anyone, anywhere on earth, at the speed of light.

And one single technological innovation, the electric light, had revolutionized the world, freeing humankind from the tyranny of darkness and night.

Thousands of inventions had sprung from humanity's mastery of the scientifically discovered principles of physical Universe, and many more discoveries lay in the future. Fuller realized, to paraphrase another great thinker of his era, J. B. S. Haldane,* that man's technological potential is not only greater than we imagine, but greater than we *can* imagine.

Fuller also knew that the earth held more than enough raw materials and resources to provide for humanity's basic needs so long as the resources were well managed. Shortages were illusory, the products of outmoded systems of thought and politics inadequate to the realities of the new world picture. There was no longer any need for anyone to struggle to "earn a living." Life was a gift, on a world blessed with an abundance of supplies. There was no longer any reason to fear, provided humanity would only accept the realization of this new truth.

Bucky's thought went one step further that cold Chicago evening.

"I had to make a working assumption. I saw that the hydrogen atom didn't have to earn a living. In fact, humans were the only ones with this game of 'You have to earn a living.' So I thought that it could be that if I was doing what nature wanted done, that I might find myself being taken care of by nature, just as the flowers got taken care of by the honeybee.

"And that would be my test, my experimental evidence, of my having absolute faith that there was a Greater Intellect—that if I were doing what the Greater Intellect wanted me to do, I would find myself being supported without me having to take any thought about it. There'd be nobody to grade my paper, nobody to tell me to do it. I'd just have to launch into it. So I had to make an experiment whether this can be done. I thought to myself, If you're doing what the Universe wants done—and let's

* John Burdon Sanderson Haldane, 1892–1964, British geneticist and philosopher.

call it God, because it's certainly not a man—if you're doing what the Greater Intellectual Integrity wants done, you'll be getting on. And if you're not getting on, then you're not doing the right things."

The last element was in place. The old Fuller was gone. In his place was a new man, a man who considered himself nothing more than an average human who had simply decided to see just what an average human could do if he laid aside his own sense of fear and limitations and devoted his life to serving others.

In the years that followed, Fuller was to become one of the most original and brilliant of the thinkers, poets, designers, inventors, philosophers, and visionaries the world has yet seen. It is no wonder that he has often been compared to Leonardo da Vinci, that great Renaissance man of all passions.*

Fuller said he had abandoned all thought of What's in it for me, and his life bore testament to his words. He adopted one single purpose: to employ all the means within his experience for the betterment of others. The light of his example continues to inspire countless thousands around the globe. He circled the world fifty times, advising presidents and prime ministers, consulting with corporate boards, addressing small gatherings of artists and young people. And his life proved a living testimonial to his grand experiment. Universe did support his endeavors. However haltingly at first, he discovered that his own needs and those of his family were progressively met as he held in mind his purpose to work only for the good of all.

Fuller's world was incredibly busy, but invariably joyous. Smiles and laughter came easily. He often adopted the role of clown and outrageous bard to cheer friends and fellow workers.

* Leonardo da Vinci, 1452–1519, painter, scientist, architect, sculptor, engineer, and mathematician; the prototypical "Renaissance man" whose life defined the standard of the diversity and genius of which one human can be capable.

He invented an amazing array of technological contrivances, ranging from the geodesic dome (which encloses more space more efficiently than any structure previously designed by humanity), to his cars, houses, chairs, roofing trusses, rowing machines . . . the list is awe-inspiring. He wrote books, eighteen of them, edited magazines, authored poetry, taught. His drawings are works of art in themselves, sketches employing the elegant simplicity, the synergy, that was at the core of his vision.

But it is as an exemplar that Fuller may be best remembered in ages to come (for his life will surely stand before the ages, if humanity transcends the impending crisis he saw so clearly). Fuller demonstrated something of the grandeur and beauty of what it can mean to be human. He confronted the fear and apathy of an age and emerged triumphant, gloriously so.

Fuller did not believe that everyone must reach the same cumulative pain and misery that overwhelmed him that night on the shore of Lake Michigan. Bucky was convinced that the transformation that overwhelmed him that day led him to a condition that is the normal and natural state of all humanity, a *birthright.* Fuller often said that his "patron," the impetus of all his creative efforts, was the children of tomorrow. For it was his most profound conviction that if a child is simply provided with accurate information in an atmosphere free of frustration that he or she would never need to suffer as he did.

That is the reason for this book.

Through the pages that follow you will accompany Bucky and three young friends as they explore the world of his thought. The journey is a friendly one, opening up new vistas and experiences and laying benign siege to much of what passes for "knowledge" in today's schools. You may find much to question; remember only that these concepts worked for Bucky, enabling him to revolutionize our understanding of Spaceship Earth and the great Universe of which it is part.

PART THREE
Fuller's Earth: Basic Bucky

What follows is Bucky's summary of the basic information he believes children should understand in order to meet the challenges of the last decades of the twentieth century.

This chapter is essentially a transcription of the first session between Bucky and the children (with additions, when relevant, from the second and third meetings). It began at the dining table of Fuller's home in Pacific Palisades, California. He seated the three children at the circular table, left momentarily, then returned with a bag full of sticks, rubber tubes, and dacron strings. As he spoke these were assembled into different forms to illustrate the geometry he presented.

The setting was intimate, to the children's scale. The children were nervous at first, but Bucky soon won them over. His victory was complete when he was able to prove that there's no such thing as a square.

FULLER: In preparation for today's meeting, I've done a whole lot of rethinking about my experiences—and I've had more than eighty-five years' worth of experiences. And I've thought especially about my experiences with education.

First of all, I should say that while I had lots of formal education in schools, I've really had more educational experiences out of school than in school, and a lot of these experiences taught me that what I had been learning in school was wrong.

I learned a lot of things in school that bothered me. But I also learned quickly that, because the teachers who were telling me these disturbing things were also the same people who were giving me my grades, if I were going to get by in school, I would have to give *their* answers, regardless of what I felt to be true. But even though I gave their answers, I didn't stop thinking; and these things continued to bother me a great deal. It's some of these things I want to share with you today.

So let's get a pencil and some paper and begin.

My teachers would talk to me about *geometry*. That's the science usually defined as the study of the mathematical properties and relations of lines, angles [where two lines meet], planes [imaginary totally flat areas reaching out to "infinity"], and solids [volumes completely enclosed by surfaces].

Now, for example, a teacher would go to the blackboard and draw a square:

A square, we were told, is a closed line with four equal angles and four edges of equal lengths. Then there was the equilateral triangle:

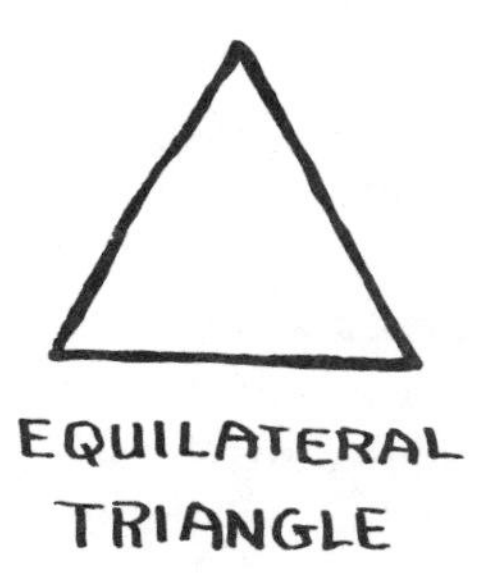

which they said was a closed line with three equal edges and three equal angles. So we were taught to look at these forms as areas enclosed by one "closed line," each formed by a line turning in on itself by angles to meet itself.

The teachers would talk about a "line" as something that went on to "infinity," unless it was bent, angled, back on itself; in which case it would form a square, triangle, circle, or what have you.

Now this one teacher who talked about the line going to infinity drew a "line" on the blackboard as she was talking. Now when I looked at the blackboard, I saw a "line" that began and ended on the blackboard. It certainly didn't go on to any "infinity." Now let me say here that in science we don't consider anything to be a "fact" or truth until we have proven it by repeated experiments. So when my teacher started talking about "infinity" again, I asked her, "Have you ever *been* there—to infinity, I mean?"

And she said, "No."

So I said, pointing to one end of her line on the blackboard, "Well, if *this* end goes to infinity, where does the *other* end go?"

And she answered, "Why, to infinity too."

Then I said, "Well then, which way is infinity?"

Then I pointed out that all she really had was a chalk line on a blackboard. And while she *said* it was a "straight line," if you looked closely, you'd see it was quite crooked; there was a lot of weaving of the chalk as she drew. (Just look at any "straight line" you draw through a magnifying glass and you will see what I mean.)

"You must get the spirit of it," she then told me. Pointing to her blackboard, she said, "This *represents* a straight line."

But I kept on pointing out that it wasn't straight. And I said I thought it was nonsense about it going "out to infinity." The fact was, the line was on the blackboard, and the blackboard didn't go out to infinity. Where did the blackboard go? Well, the blackboard was just a blackboard, a slab of the rock called slate that had been trimmed down to size. The blackboard had two large flat sides, but there were also four much thinner sides between them:

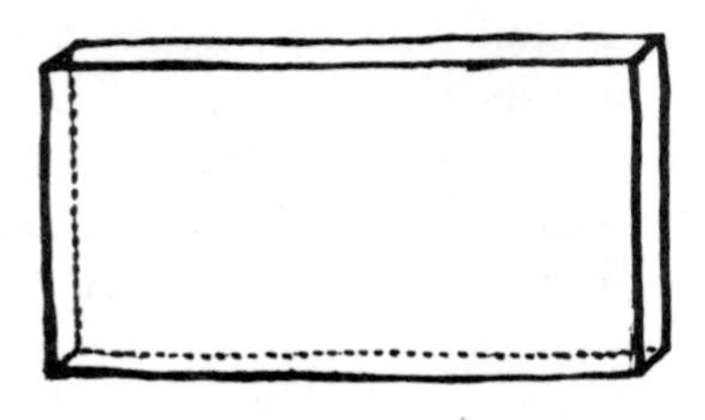

Blackboard

And if you drew a line on the blackboard and kept going, it would just go around the blackboard. Not out to "infinity."

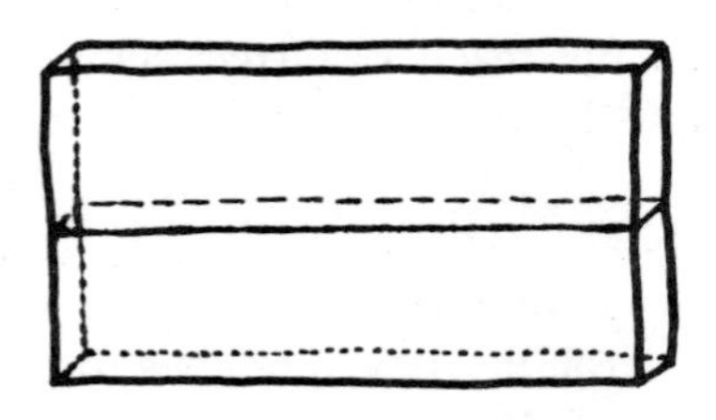

"Straight line"
goes around
blackboard

Now let's pretend you could send a line out into space like the teacher said. The first fact we have to consider is that the line would begin here on earth. Now the earth is a roughly spherical—round—planet that is revolving around its own axis at a thousand miles an hour. That means that any line you started from earth wouldn't be straight at all. It would leave a trail like a spiral or corkscrew, because the earth is also hurtling around the sun at the speed of 60,000 miles an hour.

So if we looked at our supposedly straight line from outer space, from above the North Pole, our line would look like this:

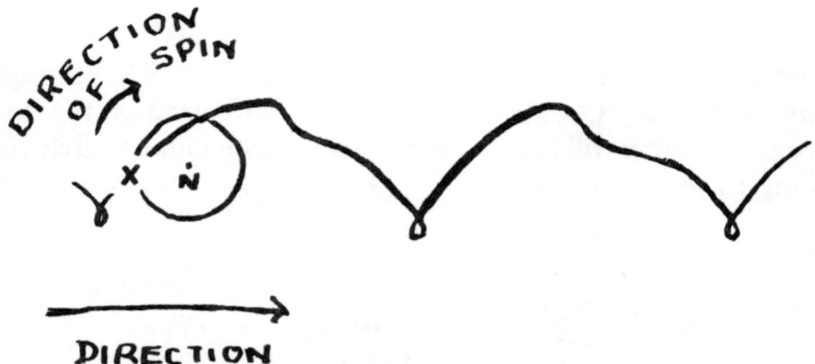

Hardly a "straight line."

So I thought to myself, Well, this teacher is certainly a nice lady, and she seems to mean well, but I don't think she knows what she's talking about. So I shut up for the time being and gave her the answers she wanted. But I kept right on thinking about what I was seeing for myself.

Now we all live on earth. It's not flat, but shaped like a ball, a sphere, and it revolves around an axis which goes right through its center. Now this axis has a North Pole and a South Pole. If we draw a line on the surface of the earth between these two poles exactly the same distance from both and perpendicular—90 °—to the axis, we will have a closed line—a line that meets itself—that we call the "Equator," dividing the earth into the Northern Hemisphere and the Southern Hemisphere. Okay?

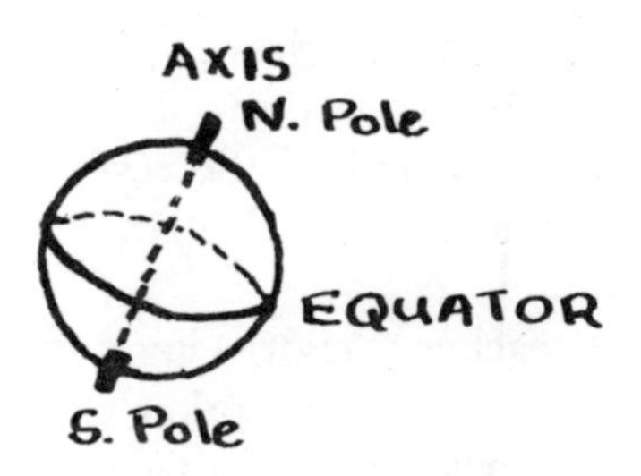

You can only draw a line on *something*, right? And whatever *something* you draw on is a shape that comes back on itself, whether it's a piece of paper with *very* thin edges, a blackboard with thicker edges, or the earth. So any line we draw on the surface of the earth will eventually become a closed line, a circle dividing the earth into two parts we call hemispheres.

Equatorial Plane
dividing Earth
into 2 hemispheres

Now let's draw another type of "closed line" on the surface of the earth—a triangle.

The earth is what is sometimes called a "closed system." In fact, all systems are closed. That is, they have a limited surface area, enclose a fixed volume of space, and are defined by a finite—limited—shape.

Now this triangle (A) which we've drawn on the surface of the earth divides the surface into two areas: the limited, measurable area inside the lines, and all the rest of the earth's surface outside the lines—an area that is also limited and measurable, although much larger.

But when the teacher was drawing a triangle on her blackboard, she wasn't paying any attention to the area outside the lines, because she was talking about this nonsense of infinity. But her blackboard is just as much a "closed system" as the earth. It has a limited surface, encloses a definite volume of space, and has a definite shape.

And this brings us to an important point. If you make a

closed line on any system, it divides the surface of the system into two areas: one inside the closed line and the other outside.

But the teacher was saying I could only describe one side of the line, the area "inside."

So we have all these school people getting you to look only at the little things, instead of the big things.

I discovered that if I made a triangle on the surface of the earth, I have divided the earth into two areas. Now the area inside the triangle I can define as an area bound by a closed line composed of three angles and three edges. *But I also find that the area "outside" is also bound by a closed line of three angles and three edges.*

Let's say that each of the angles in triangle (A), the "inside" triangle, is approximately 60 °. That would make our triangle (A) an equilateral triangle. The sum of all the angles— 60 + 60 + 60— would be 180 °. But then I look at triangle (B), our outside triangle, and I see that here I have three angles of 300 ° each, for a total of 300 + 300 + 300, or 900 °.

Again, I find that school is always eliminating and pushing out of the way everything that is really big, the really big things.

Now the teacher would say, "Well, when I draw that triangle, I didn't *mean* to divide the earth into two areas." But the *fact* is, that's exactly what she *did*. So, you must see that when you thought you were only doing little things, you were really doing very big things. It's very important for us to realize that. Not only do we divide something *this* way; we also divide it *that* way. So any time you do something like this in Universe—take a little bit out here—you are also creating a very big "rest of Universe" there, like our triangles (A) and (B). You have to pay attention to what you're really doing, because in school they make you look at things oversimply.

Does that make sense to you?

JONATHAN NESMITH: Yes, very much so.

FULLER: Good. Now let's look at some other triangles as they really exist on this earth.

First, let's draw our earth again, with its axis, its North and South Poles, and the Equator.

Now I want to tell you about something called a "great circle." That's a term meaning any line formed on the surface of a sphere by a plane running through the center of the sphere. A great circle is formed by a plane that slices the earth exactly in half. The Equator is a great circle. So are all the meridians of longitude: they are the great circles that pass through the two poles, as well as the center of the earth, and are used in navigation to "fix" locations.

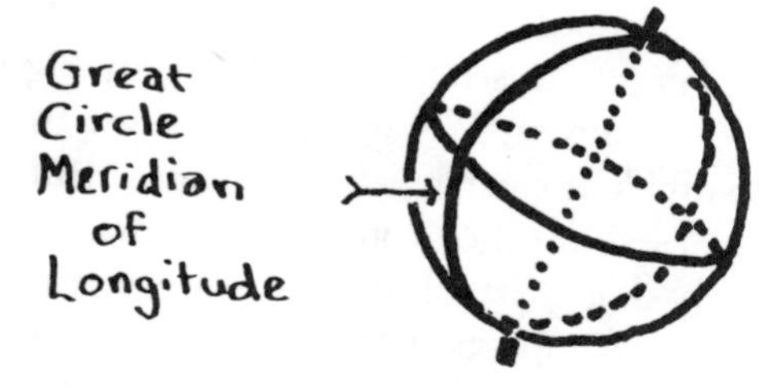

The Equator is the only great circle that passes through the center exactly perpendicular to the axis.

But the meridians of longitude and the Equator aren't the only great circles (these are just the ones most commonly used in mapmaking). There can be any number of great circles drawn that pass through the center of the earth exactly dividing the planet into two equal halves but not passing through nor perpendicular to the poles. Here are a few examples:

Now there are also circles we call "lesser circles." All circles drawn on the surface of a sphere that are not formed by planes passing through the center of the sphere are all called "lesser circles." All lesser circles are smaller than great circles drawn on the same sphere; the great circle is the largest possible circle that can be drawn on the surface of a sphere.

The most commonly known and used lesser circles are the lines of latitude. The Equator is also a line of latitude, but the only one that is also a great circle. Lines of latitude are all the lesser circles and the one great circle Equator that are perpendicular to the axis of the earth, but with the exception of the Equator do not pass through the center of the earth. Lines of latitude are described in terms of north or south. They are either parallel to and north of the Equator—between the Equator and the North Pole—or parallel to and south of the Equator—between the Equator and the South Pole.

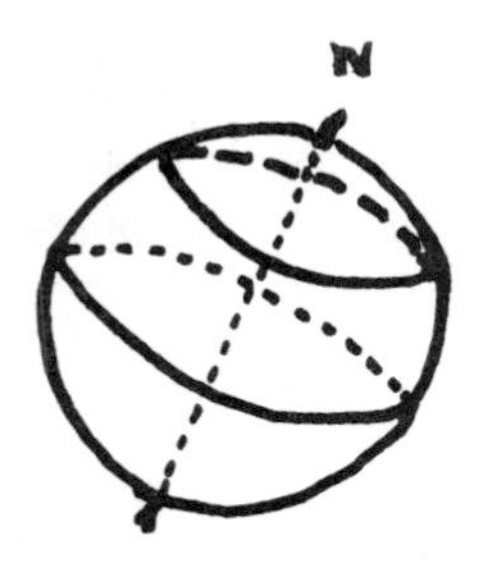

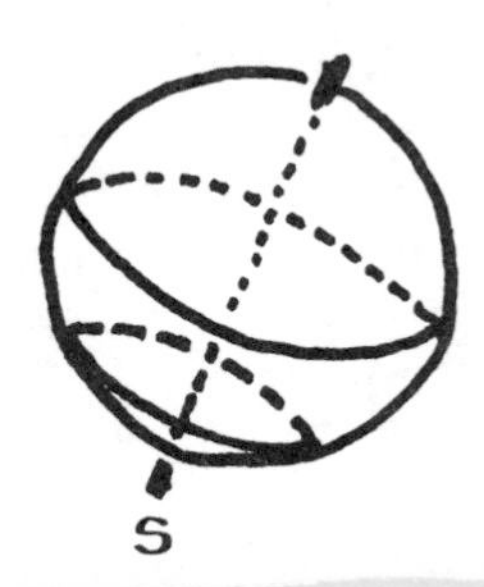

The lines of latitude are measured in terms of degrees. These degrees are calculated by measuring the angle formed from any point on the latitude line (A) drawn directly to the center of the earth (B), and from there back to a point on the Equator directly above or below the starting point (C) [that is, on the same meridian of longitude as the starting point].

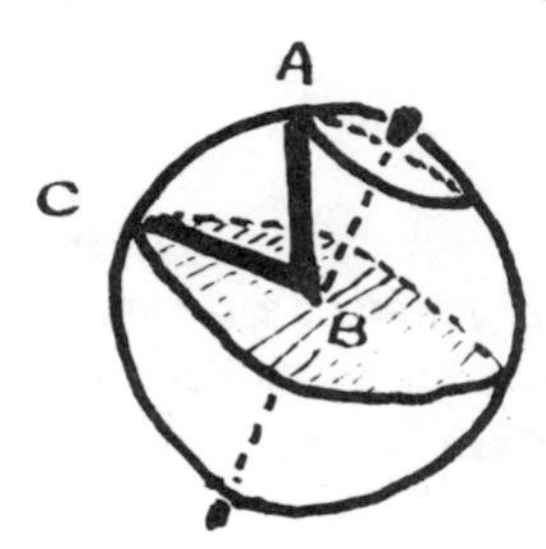

The North Pole is 90 ° north latitude; the South Pole is 90 ° south latitude. A point midway between the Equator and the North Pole would be 45 ° north latitude; a point midway between the Equator and the South Pole would be 45 ° south latitude.

So when we talk about longitude and latitude, we are talking about great and lesser circles drawn on our spherical earth. All lines of longitude are great circles; all lines of latitude are lesser circles, except the Equator.

Now I'll take a pair of dividers—a compass—and mark off where 80 ° north latitude would be up here by the North Pole.

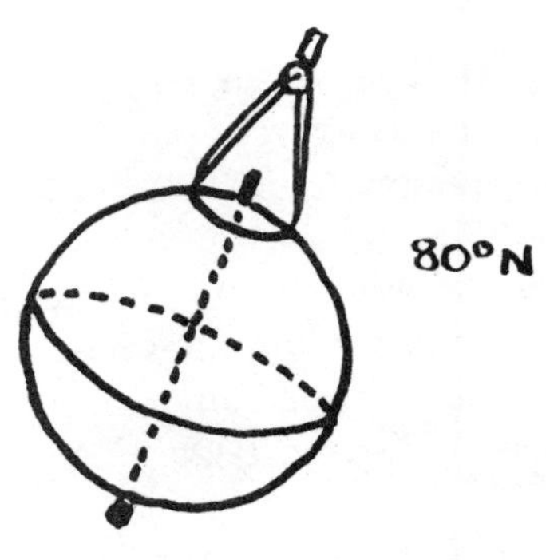

Now that my compass is set for a lesser circle exactly the same size as the circle at 80 ° north, I'll take it and put the sharp point anywhere along the Equator and swing the pencil side to create a circle of the same size. There.

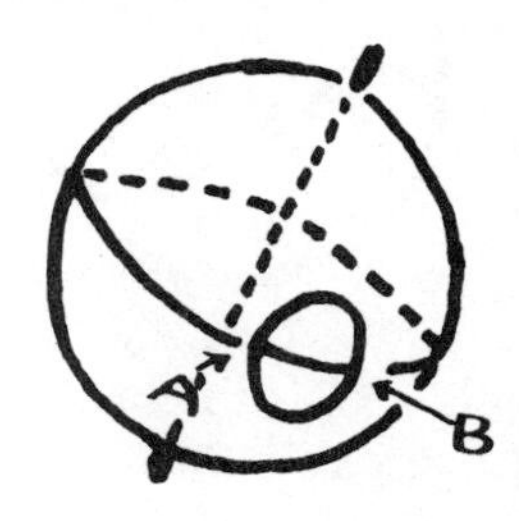

Now we can see that this lesser circle crosses the Equator at point (A) and again at point (B). You can also see that it is much shorter to get from (A) to (B) by staying on the great circle Equator than by taking the lesser circle. So we find that great circles are always the shortest distances between any two points on the surface of a sphere. That's why ships have always tried to follow great circle routes when navigating great stretches of ocean. Do you follow?

BENJAMIN MACK: Yes, I see.

FULLER: Now what I have been doing these past few minutes is introducing you to what is called "spherical trigonometry." "Trigonometry" is simply a word that means the study of the relationships of lines and angles in the triangle. Spherical trigonometry is the study of triangles formed on the surfaces of spheres. All navigation of ships and airplanes uses spherical trigonometry.

The circles we are using in our spherical trigonometry are the equivalents of the straight lines on a plane. A straight line is the shortest distance between any two points on a plane; a great circle is the shortest distance between any two points on the surface of a sphere.

Now I'm going to take a meridian of longitude and bring it down from the North Pole (A) to the Equator (B):

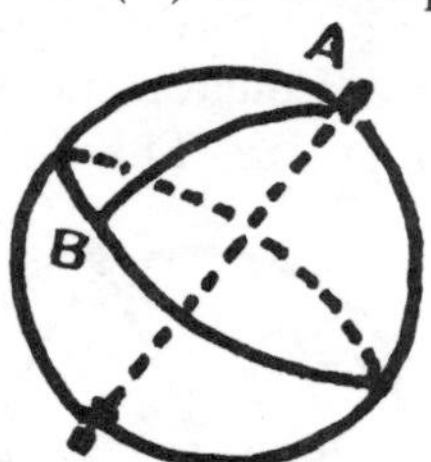

Now we already know that the Equator is inherently perpendicular—at exactly 90 °—to the axis. The Equator can also be defined as the great circle formed by the spinning of the earth around its axis.

Now our meridian (AB) is perpendicular to the Equator, as are all meridians of longitude. So the angle formed by the intersection of (AB) and the Equator is 90 °, a perpendicular angle formed between the planes of the meridian and the Equator.

Now I'm going to leave point (B) and move along the Equator one-quarter of its total length to a place I'll label (C). Now we all know that there are 360 ° in a circle, which tells us that one-quarter of a circle would be 90 °.

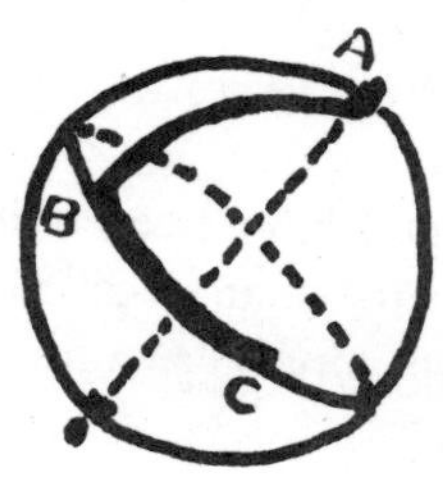

Now I'm going to run a new meridian from (C) back up to the North Pole (A). This gives us two angles: (BCA) and (CAB). We already determined that our first angle (ABC) is 90 °.

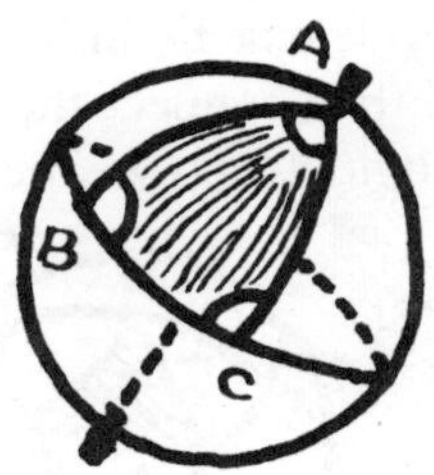

Knowing that angle (ABC) is 90 °, we can determine that angle (BCA) must also be 90 ° for the same reason: it is formed by the intersection of a meridian of longitude with the Equator. Now our angle (CAB) is also 90 °, because it is formed by the intersection of two meridians of longitude separated by exactly one-fourth the length of the Equator, a great circle of 360 °, one-fourth of which must be 90 °.

Now let's look at what I've done. I've taken one meridian from the North Pole to the Equator (AB), moved one-quarter the length of the Equator to (C), and from there formed a new meridian to the North Pole (CA). Both (AB) and (CA) are sections of meridians of longitude, and therefore sections of great circles. Because the meridians (AB) and (CA) are great circle lines from the North Pole to the Equator, they are both the same length. We also know that the Equator is exactly midway between the two poles, and at 90 ° to the axis. Thus, the distance between the Equator and a pole is exactly one-quarter of a great circle. Now we know that all three legs of our triangle—(AB), (BC), and (CA)—are exactly the same length, one-quarter of a great circle. We have also proved that the angle formed by the intersections of each of the legs with another is 90 °.

What we have formed here is an equilateral spherical triangle with three equal legs (each one-fourth of a great circle) and three equal angles (each 90 °). 90 ° + 90 ° + 90 ° = 270 °. And that's clearly *not* the 180 ° your teacher told you every triangle must be. We have constructed a very real triangle on a very real earth, and we've come up with 270 °.

Now let's go a few steps further.

Let's bisect—divide exactly in half—each leg of our spherical 270 ° triangle and then connect the midpoints to form a new, smaller spherical triangle which we'll call (DEF).

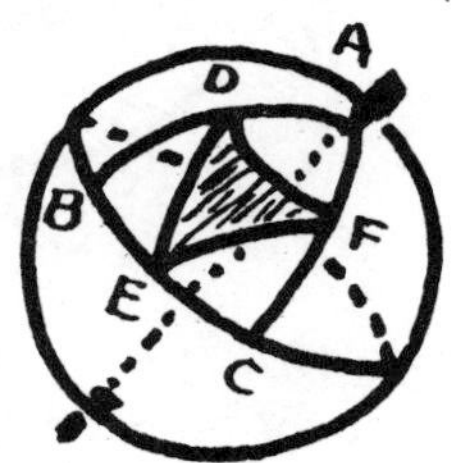

(D) is exactly midway between (A) and (B); (E) is midway between (B) and (C); (F) midway between (C) and (A). If we measure each of these three new angles—(FDE), (DEF), and (EFD)—we'll find that each measures about 73 °, for a triangle with total angles of 219 °—certainly not our 180 °.

Now we can repeat the same process with (DEF), bisecting each of its legs and connecting the midpoints to form a new triangle we'll call (GHI).

Each of the angles in our new triangle is about 63 °, for a total of 189 °. Again, no 180 ° triangle.

We can continue this process, making smaller and smaller triangles. But as close as the sum of the angles may approach 180 °, it can never exactly reach that number, *because 180 ° can happen only on an absolutely flat plane, and there's no such thing as an absolute plane in the universe.* So no triangle ever adds up to exactly 180°. It will always be more. A triangle can approach 180°, but it can never reach it.

This is the way you ought to be taught. They make the mistake in school right from the beginning by oversimplifying, and by saying that spherical trigonometry is too complicated for you. So they say, "I'm going to give you *plane* geometry," not even cubical geometry. So you have to pretend there's something called a "plane," even though you can't have a surface by itself—it has to be the surface of something. And anything that has a surface must also have an *insideness* and an *outsideness.*

So you might as well start with reality, and not with the fake imaginary plane that doesn't even exist.

They tell you, "Well I'm going to start with something simple—a *point* that really doesn't exist. It's an *imaginary* point. Now I take a row of these points, and that's one dimension, a *line.*" But you've already told me a point is nothing, so how can you make a line out of "nothing"? And besides that, *it's impossible to have just one dimension by itself.*

If you're a scientist, you can't accept anything without experimental evidence, and there's no experiment that's ever been devised to prove the existence of just one dimension by itself. Now you can have the linear dimension of a *polyhedron* of some kind. A polyhedron is an object having several surfaces and enclosed space; *poly* is Greek for "many" and *hedron* means "face" or "surface." You can't have a *line* by itself; you can't have a plane by itself; you can't have a surface by itself. You can't even have a cube by itself unless it has weight, unless it has longevity, unless it has a temperature—all these qualities of existence.

JONATHAN: What do you mean by longevity?

FULLER: How long it has existed, how old it is. Anything that exists must have these qualities.

I want you all to be scientists, pure scientists right here from the beginning. You have to ask these questions: How big is it? How old is it? And so forth.

I know it can be very hard for you, having worked so hard at school, to discover that what you've learned is wrong. But I'm giving you some corrections now. So let's talk a little more about lines.

Now my teacher had a very nice ruler, a very good steel one. And she used that to draw her next line. But I told her, "Your line is *still* crooked. You can get a magnifying glass and really see how crooked it is. The chalk is uneven, and the line simply isn't straight."

But the teacher said, "You're simply not getting into the spirit of mathematics. This is a straight line."

"It is not," I said.

"Very well, then," she said. "I mean a *line of sight,* as when you are looking directly at something far away."

But let's suppose I get a telescope and make a "line of sight." Let's put the telescope so it's pointing directly at the point where the sun is tangent—touching—the horizon in the evening, just before we lose sight of it.

Now the sun is 93 million miles from the earth, and it takes light, traveling at 186,000 miles a second, about eight minutes to reach the earth. That means that the sun really hasn't been where our telescope is pointing for eight minutes. We're actually seeing *around the curvature of the earth,* and that's *not* a straight line.

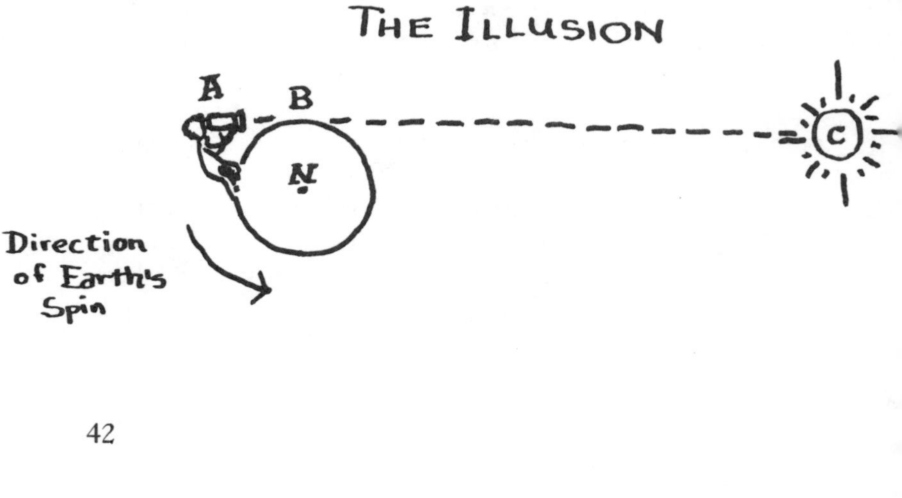

THE ILLUSION

A
B
N
C
Direction of Earth's Spin

Our observer is standing on the Equator, and we are looking at him from a point above the North Pole. What appears to our observer (A) as he looks at the sun touching the horizon (B) is that he is seeing in a straight line from himself to the horizon and on past to the sun. But the earth has already revolved some eight minutes further than when the light left the sun. So our observer is actually hidden from the sun's real position at the moment he is seeing light eight minutes old.

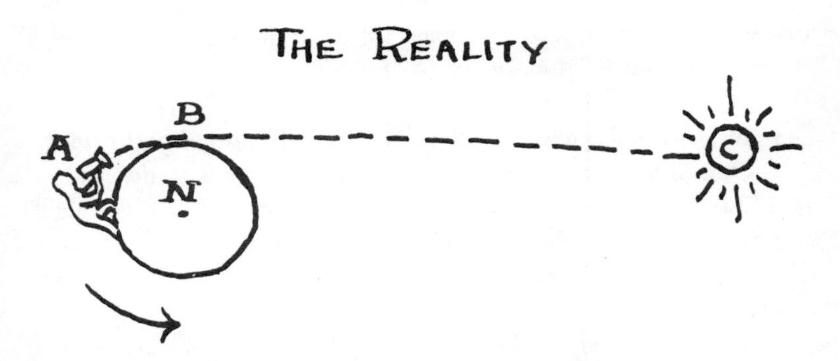

The observer is actually *seeing around* the curvature of the earth. This is what Einstein* meant when he talked about "curved space."

So here was this teacher telling me I wasn't getting into the "spirit of mathematics." But the "spirit of mathematics" she was talking about was to make up things like "The sum of the angles of a triangle is always 180 °" and "A triangle exists all by itself, just on *one* side of the line." And that's why we have this immediate bias: "I'm only interested in this side of the line. You can't pay any attention to the other side, because it goes all the way out to infinity." But I said it didn't go out to infinity at all.

* Albert Einstein, 1879–1955, a German-born physicist who revolutionized the sciences with his formulations of the general and special theories of relativity.

I'm interested in the *rest of the world,* and that's what's on the other side of the line, *not* infinity.

So I said to my teacher, "I'm going to be a mathematician."

One of the greatest of the mathematicians was a man named Boole.* Now Boole found that when you couldn't get an answer, the best approach was "Just be completely absurd—make up the most absurd (ridiculous) answer you can think of. Make up a deliberately absurd answer. Then if it's really absurd, then I can get a little less absurd, and then a little less absurd, and then still less absurd, until finally I may get somewhere near the area where I'll be correct. At least I'll be in the right area." This process is called "reduction from absurdity."

Now I'm going to do a beautiful thing in the spirit of Boole. I'm going to make what I call a "deliberately non-straight line."

One of your ways of defining a straight line is to describe it as a line whose ends never come back on themselves. So we'll begin our deliberately non-straight line with a length of rope whose ends we'll splice together. So we're starting with a line whose ends come back on themselves.

Now we'll use a Dacron rope, because Dacron doesn't stretch; it always remains the same length, unlike most other materials. And when we look closely at the rope, we see that it is made up of a lot of individual strands that are all woven around each other in curly spirals. So not only does our line come back on itself, but it's composed of fibers that curl around each other.

Now once I've spliced the ends together, I've made a loop that's coming back on itself.

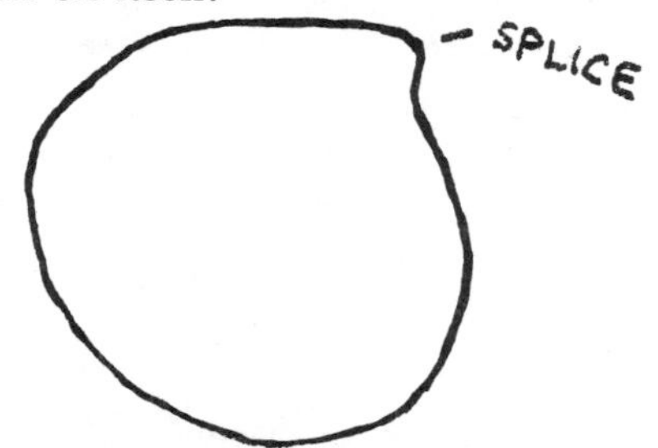

* George Boole, 1815–1864, an English mathematician and scholar who specialized in the application of mathematics to logic.

Next, I'm going to take any two parts of the loop and join them together in my hand.

Now I'll put a clamp where they come together, and then I'll massage along the rope until I reach a point where the two parts turn around. And now I'll tie a red ribbon at that point.

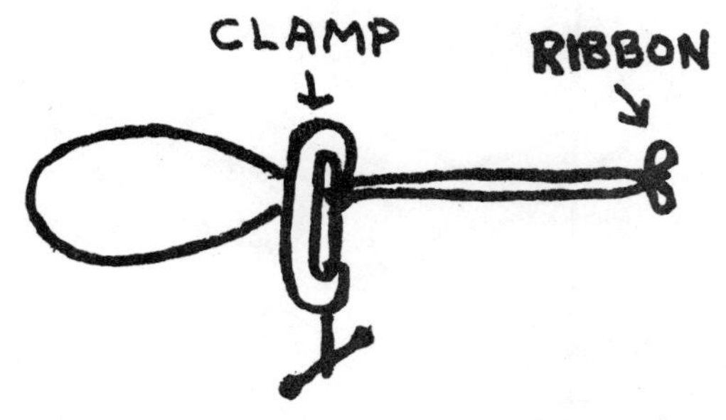

Now I'll go back to the clamp and massage the rope in the other direction until I reach the turnaround point, and I'll put another ribbon there.

Now a circle is a finite [limited] phenomenon. It comes back on itself; it doesn't go out to infinity like the imaginary line.

So what we've done with our red ribbons is to divide our rope into two halves, two equal lengths. The ribbons mark the halfway points.

What I'll do next is take our rope and bring the two red ribbons together, and then massage the two lengths of rope together until I reach their turnaround points, where I'll tie two blue ribbons.

Now we have halved our two halves; we've divided our rope into quarters. We can keep on with this process, dividing the rope into eighths, sixteenths, thirty-seconds, sixty-fourths, and so on. I can make as many divisions as we want.

Next, I'm going to a large wall, and we're going to drive two nails into the wall at approximately the same height from the floor and at a distance apart from each other less than the distance between the two red ribbons. Now we'll loop our rope over the two nails.

Now with the help of two friends, we'll stretch the rope above and below the two nails until it's tight, holding the rope at the blue ribbons which divide it into fourths. And when the rope is tight, we'll drive in two more nails where the blue ribbons are.

This creates what we call a rhombus, a shape of four equal edges and two pairs of angles, each pair equal to itself but not to the other pair.

Now let's say that when we were dividing our rope into successively smaller halves that we marked our next division—that would be eighths—with green ribbons.

Now we'll put nails inside our rhombus just touching each of the green ribbons. Next, we'll pull the rope off the two nails where the blue ribbons are and bring the blue ribbons together so that they touch. This results in two rhombuses exactly half the size of the first.

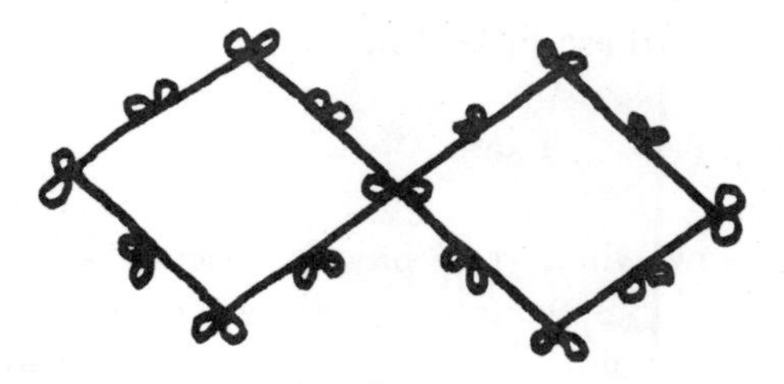

Now we can repeat this same process with the next divisions, the sixteenths. This gives us *four* rhombuses, each one-quarter the size of the first. Each of the rhombuses has four sides, for a total of sixteen—and this was the result of our dividing the rope into sixteenths:

Now we can repeat the process again with our next division, the thirty-seconds. This gives us eight diamonds.

And if we do it one more time, with the sixty-fourths, we get sixteen diamonds (rhombuses).

You can see that we won't have to continue this very much longer until we get something that *looks* like a straight line. But you and I know that it's not a straight line. It's a deliberately non-straight line.

JONATHAN: Right, I understand.

FULLER: So what would look like a straight line is simply very, very many of these little diamonds.

BENJAMIN: Yes . . . I see.

FULLER: Now when your physics teacher wants to teach you about a wave, like electromagnetic waves and so forth, he takes a piece of rope and nails one end to the wall and takes the other end in his hand. Next, he pulls the rope tight and whips it. When he does this, you see a wave go to the wall and then come back to his hand. You can see this for yourself by tying a rope to a door handle. The important thing to note is that the wave makes a complete loop every time.

What is happening with our deliberately non-straight line is like an electromagnetic wave. And when a mathematician tells you, "I mean a line of sight," you must remember that sight is like a magnetic wave: very, very high in frequency—number of complete loops per second—and very, very small in wave length—the distance between the starting and finishing points of one complete loop cycle.

A low-frequency wave is like our one-rhombus "line"; and a high-frequency wave is one that would *appear* straight but is really not. But a line of sight is a *wave* line, "wavilinear," like our deliberately non-straight line, and not a straight line. *Physics has found no straight lines, only waves.* So we're in the world of reality when we refuse to accept this "reality of a straight line," because there's no such thing.

BENJAMIN: So there're no straight lines, ever?

FULLER: That's correct. How can you make a straight line in a world where you're on the surface of a sphere that is traveling a thousand miles an hour around its own axis, orbiting the sun at 60,000 miles a hour, and the sun and all its planets are circling the center of the galaxy at an even faster rate of speed, and the whole galaxy is moving through space at a still higher speed? So with all these aspects of Universe in all this motion, where are the straight lines?

I find it wonderful how much you're learning and how quickly you catch on to what the truth is. What you can prove for yourselves as pure scientists is fine—but don't get fooled by what they teach you in school as absolutes.

Today you've been learning about *relationships.* And I hope you can see that you must learn to be logical, and to rely on experimental evidence, evidence which can be proven just as we have been proving things here today. Experimental evidence is evidence about the universe we live in, which is the reality.

One more thing about lines. Every line is a history. All lines are the consequences of some action, whether it's done by you with a pencil, or by some star spinning by you. A line is a history, and the front end of the line is the event creating the history, like the point of the pencil moving across the paper.

Now let's go back to the classroom again. The teacher went

to the blackboard and drew a square. But the only reason her square stayed square was because it was held in position by the blackboard. A square can't hold its shape at all. In fact, there's no experimental evidence to prove there's any such thing as a "square" at all.

We define a "plane" as a surface formed by three points [one point is a point, two points form a line]. But if we have four points, we have formed a *hinge,* and a hinge doesn't have any stability at all. This is very easy to prove for ourselves.

We'll take four sticks of equal lengths and join them together with flexible rubber connections. It takes four connectors to join the four sticks. Now you'll find that the form we have created has no stability, no ability to hold its own shape. If you pick it up, it wobbles every which way and certainly doesn't look like a square.

But if you take *three* sticks and *three* connectors, you'll create a form that does hold its shape, a triangle. The triangle is the only polygon—the only "flat" figure we can create with sticks and connectors—that will hold its own shape.

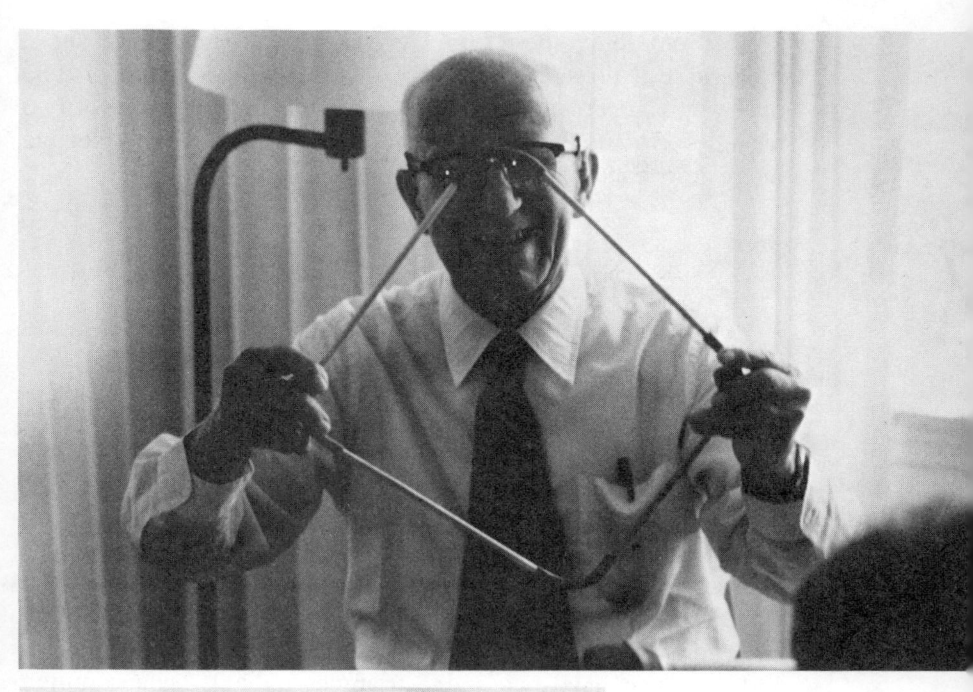

If something doesn't hold its own shape, we can't talk about it, because it doesn't exist. When is a non-shape a shape? The only way you could make a stable square is to form it out of two triangles. Because only in the triangle is every potentially flexible hinge restrained by an opposing push-pull bar.

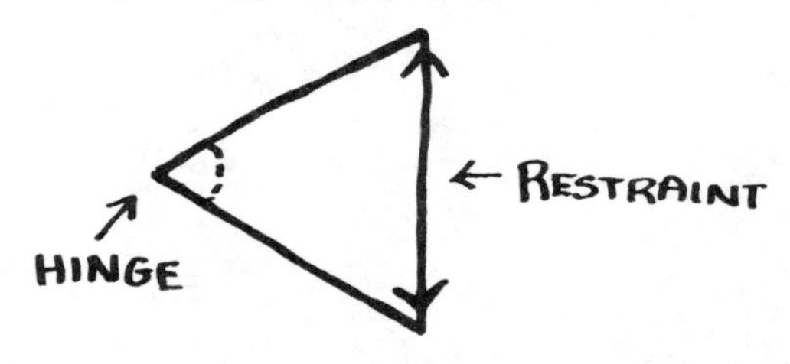

BENJAMIN: So you're saying that the only shape that really exists is the triangle?

FULLER: Yes. Now we've already seen that a square can't hold its shape. It has no stability.

You have to add another stick to get any stability at all. You can run a stick between any pair of opposite angles in a square, and now you have two stable triangles moving on a common hinge, our new stick.

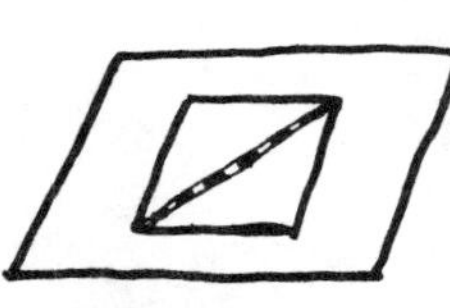

Square with "brace" sitting on flat surface becomes . . .

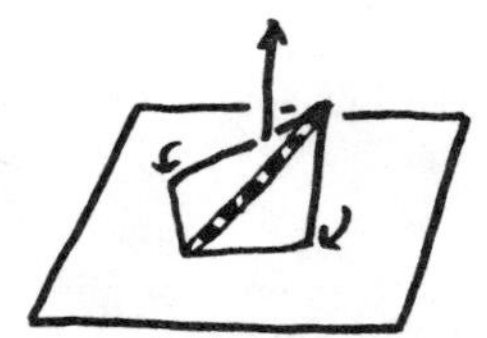

. . . two triangles moving together on a common hinge as brace is pulled away from surface,

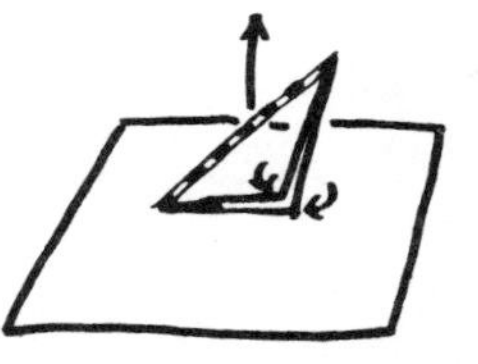

. . . becoming two congruent (touching) triangle with one shared edge.

Again, the only thing that will hold its shape is a triangle. So when you're talking about real structures, you have to start with the triangle.

Now how does a triangle hold its shape? You remember learning in school about the lever? Like when you take a screwdriver and pry open a paint can? A lever enables you to exert force far stronger than you could simply using your bare hands.

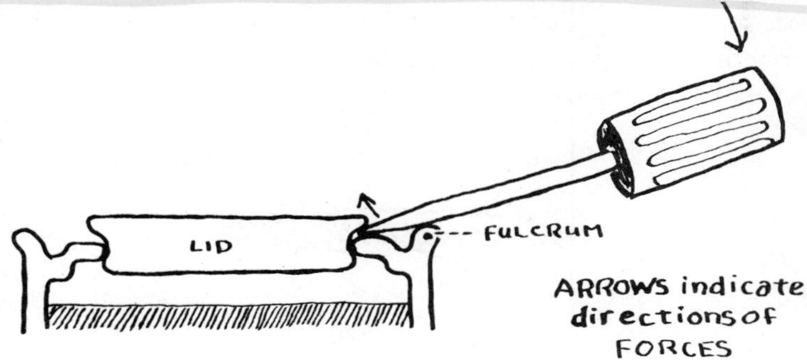

Now we call the power a lever gives "advantage." The lever uses a stable base, a fulcrum, to rest against when it exerts pressure. In our paint can example, the screwdriver is the lever and the "lip" of the can is the fulcrum. The power of the lever becomes increasingly greater the longer the distance between the fulcrum and the place where you are applying the pressure, and the shorter the distance between the fulcrum and the object you are exerting pressure against. So a lever with a short arm is much weaker than a lever with a long arm.

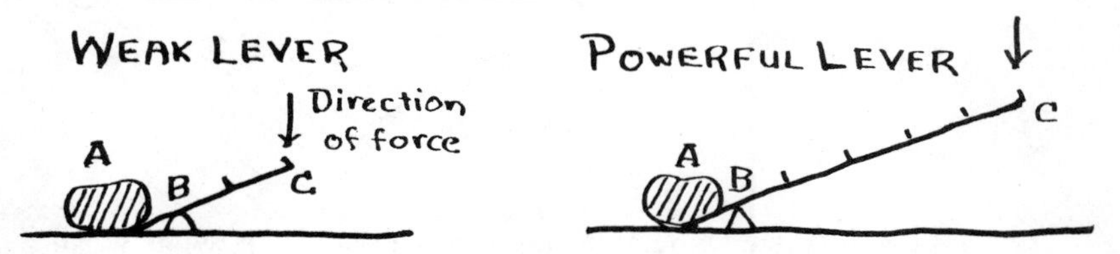

Our strong lever is much stronger than our weak lever because the distance between the source of the application of pressure (C) and the fulcrum (B) is greater, the distance between the fulcrum (B) and the object to be moved (A) remaining the same in both instances. The longer the "handle," the greater the advantage. You can prove this for yourself.

The Greek mathematician Archimedes [287–212 B.C.] once said, "Give me a lever long enough and a place to stand, and I can move the world."

Now if you take two *levers,* you can make yourself a pair of scissors. The handles are the pressure points, which you squeeze together with your hands, and the pin that holds the two pieces of metal together is the fulcrum.

Now the longer the handles in relation to the blades, the more power the scissors have, the heavier the materials they can cut. Bolt cutters, used to cut thick pieces of steel, are really very powerful scissors with very long handles and very short blades. Using this tool, the average person can exert thousands of pounds of pressure; that's how much advantage they give.

Now what we have with one angle of a triangle is a pair of levers, like a pair of scissors.

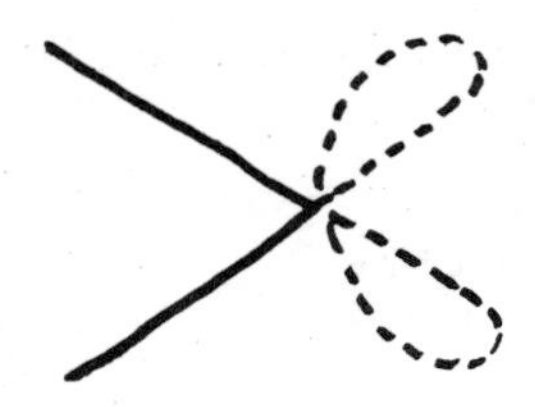

The third side, the side opposite the angle, is what I call a "push-pull." It takes hold of the opposite angle and stabilizes it. Two sticks joined together by a flexible rubber connector can flop any which way, but once they are opposed by the push-pull, they become locked into place.

So each of the three angles in a triangle is a pair of levers opposed by a push-pull. There's no other structure like it.

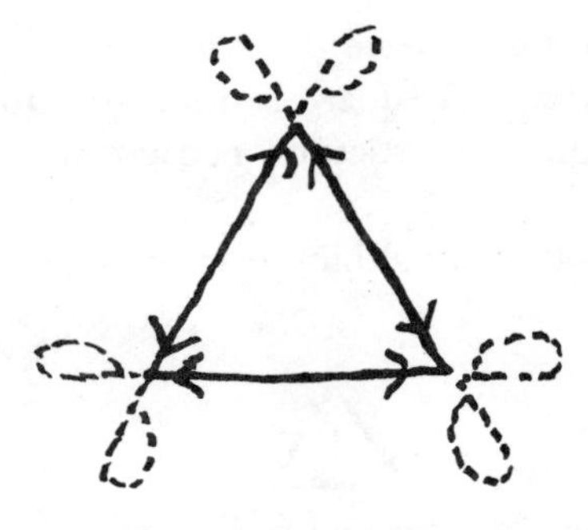

EQUILATERAL TRIANGLE as three scissor-lever angles, each opposed by a "push-pull"

What I began to see as I continued to look at what I was being taught compared to what I was learning to be true showed me that physics and engineering didn't have any real definition of *structure.* They said, "A structure is obvious; a block of marble holds it own shape." But I said, "Wait. You don't know what the atoms are doing *inside* that block of marble, and that's what you have to look at."

So when I use the word "structure," what I mean is "a complex of events."

This rubber piece connecting our two sticks at the corner of our triangle has a lot of atomic events going on inside it. And that's also true with our sticks; each one is a complex of many atomic events [just as a house can be seen as a complex of thousands of bricks, boards, and nails].

So our one triangle is also composed of six major parts: our three flexible tension corners and our three rigid push-pull edges. And these six interact with each other to produce a stable balance.

So, I say that *a structure is a complex of events that interact to produce a stable pattern.* A square is not a structure, because it is not stable; a triangle is a structure, because it is stable.

Now, let's look at something else. First, I'm going to draw one triangle.

This triangle represents the number 1.

Now I'll draw a second triangle, a bit bigger, and I'll divide its edges in half and connect the dividing points.

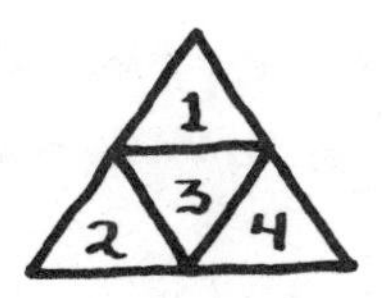

This gives us four triangles.

Now I'll draw a third triangle, still larger, and divide each of the edges into thirds, connecting the dividing points as we did before.

This gives us nine triangles.

Now let's make one more triangle, the largest yet, and divide its edges into fourths and connect the dividing points.

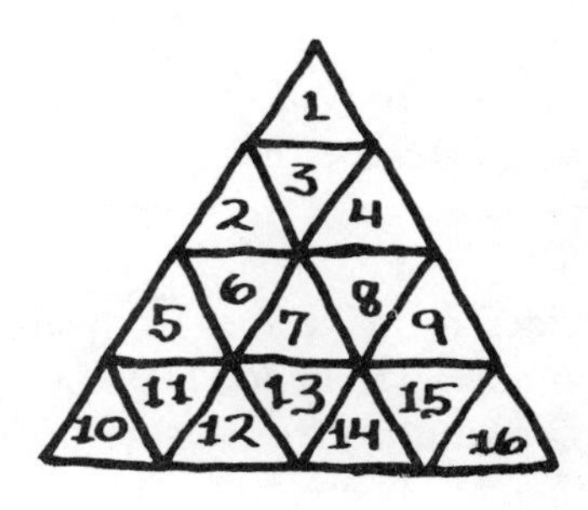

This gives us sixteen triangles.

Now you remember how they taught you about "squaring" numbers in school? 1 × 1, or 1 "squared" (1^2) equals 1; 2 × 2, or two squared (2^2) equals 4; 3 × 3, or three squared (3^2) equals 9; 4 × 4, or four squared (4^2) equals 16. And maybe they'd show you drawings like this:

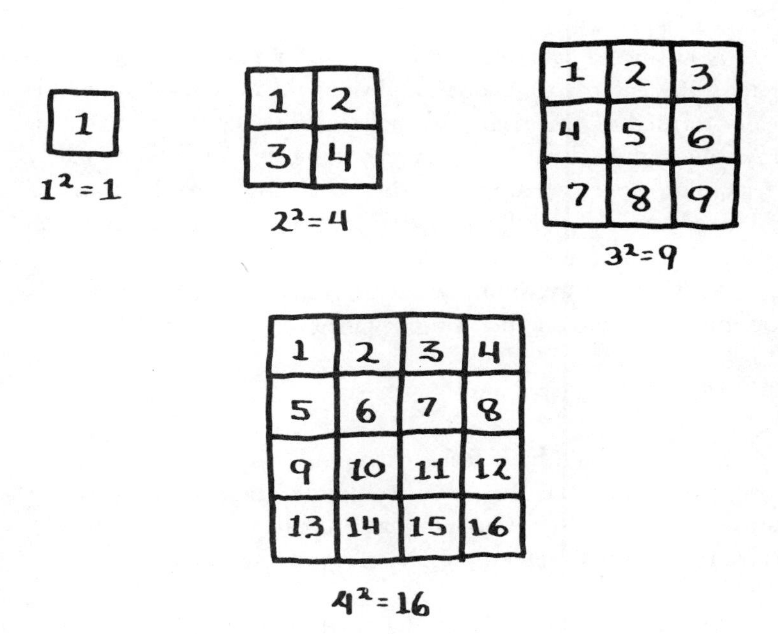

But really the most efficient way is "triangling" instead of "squaring." We've already proved that a triangle is stable, while a square is not. Now we can see that a triangle is also more efficient than a square; it does more work with less material and effort.

It takes four sides to make the 1^2 square, but only three sides to make the number-1 triangle. There are nine sides in the number-4 (2^2) triangle, and twelve sides in the "4" square, three more than in the triangle. There are eighteen connectors in the number-9 triangle, and twenty-four connectors in the "9"square. For the number-16 triangle, there are thirty connectors, and forty connectors in the "16" square.

So not only is the triangle stable where the square is not, but the non-stable square takes one-third again as much material and effort to contain the same number of units (subdivisions) as the triangle.

So what's really happening is that nature is "triangling" instead of "squaring."

And since a square has no integrity, no ability to retain its shape, whereas a triangle does, and since the only stable square is really formed of two triangles, and since we say that nature always does things in the most economical way, you're really "triangling" even though you say "squaring." And because a stabilized square is really two triangles, it takes twice as much area to "square" as to "triangle."

So when nature multiplies herself times herself, as she does continually, would she be using squaring or triangling?

BENJAMIN: Triangling!

FULLER: She'd have to be triangling, since she's always the most economical. She has to use something that works, that she can count on. We've seen you can't count on squares, and so teachers at school are completely wrong about that.

RACHEL MYROW: When I hold the square, it just keeps

going back and forth in my hand like it's trying to become a triangle.

FULLER: Right, but it will not become a triangle until we put in that one more stick that makes it two triangles. Now it tends to hold its shape.

But even now you can't guarantee that it will stay in one plane, because the two triangles hinge around each other. They only stay a square when you keep them on a flat surface like a table. There's still something else we need.

Now I want you to pay very close attention to what I'm saying now. If I can point to anything, it's because it's seeable. And it's seeable because there is light bouncing off of it back to my eye. Now if I take my pen and touch it to the paper, I may say, "I've got a point here."

POINT

The fact is, I've put ink there. I say, "I'm seeing a point." But what I'm actually seeing is light reflected off the material that makes up the ink. If you turned out all the lights and closed the curtains so that the room was completely dark, you wouldn't be seeing any point at all.

And if I looked at my point through a very big microscope, I could see that it has matter. It's composed of all the chemicals that make up the ink. So anything I see has light bouncing off its surface. You can't have a surface of nothing; it has to be a surface of *something*. And if it's a "something," it has to have an insideness and an outsideness.

So now I get to a very important question. *What is the minimum that will give me an insideness and an outsideness?*

Now two points have "between-ness." Three points also have "between-ness," but no insideness and outsideness. Three is still open, like a circle. But with four points I have an insideness and an outsideness. *Four encloses a volume of space.*

Now if you take a sledgehammer to some rocks and keep breaking them up, you'll find that you can never get a piece with *less* than four corners; and you can never get a corner with less than three edges coming together; and you can't have a face with less than three edges.

So there is one structure that is very important. It has four corners, each with three edges coming together, and it has four faces, each with three edges.

This structure is the minimum "something" in the universe.

Now the minimum "something" in the universe must be very important—and it is. It's what we call the "tetrahedron." *Tetra* is Greek for "four," and *hedron* means "sides."

But it really isn't safe to call these "sides," because there are really no sides to it. There are four triangular windows. The idea of "sides" is another old error dating back to the days when a people saw a block of marble as an absolute solid, before scientists discovered that all so-called sides are really composed of very, very tiny atoms separated from each other by relatively large amounts of space.

So we have our triangular windows.

Now the minimum "something" of the universe has four corners—we can call them "loci"—and four triangular windows. But it also has six edges connecting the four corners and forming the four windows. So there's the number 6 as well as the number 4.

So we see that the minimum something begins with a 4—the four corners and faces—but it also includes a 6—the six edges. So 4 and 6 are the minimum numbers. There is no number 1 by itself; 4 and 6 are the numbers that begin the minimum "something." They are, therefore, very important numbers.

Now there was a man named Euler,* a mathematician. And Euler said that anything you can see breaks down into three fundamental (basic) elements of see-ability.

One is a *line*. Then there is a *vertex*, the place where two or more lines meet or cross. Then, when three or more lines cross each other to delineate—outline—a space, you have an *area*. So you have the *line* (A), the *vertex* (B), and the *area* (C).

* Leonhard Euler, 1707–1783, a Swiss mathematician who wrote more than seventy books in all areas of mathematics. He has become especially noted for the formula Fuller mentions.

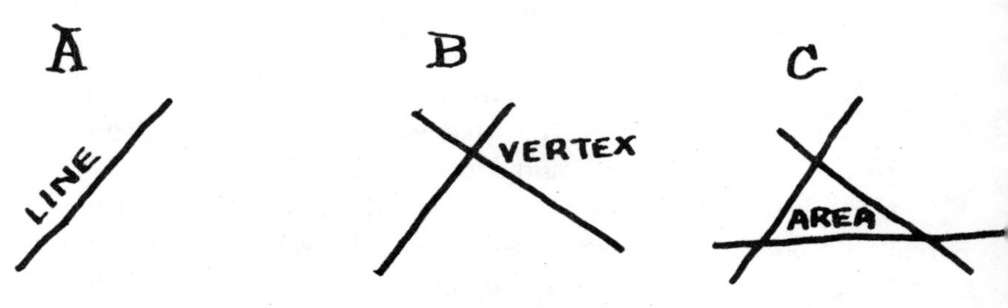

Now a vertex is formed by two or more lines crossing or intersecting at the same point. An area is formed by three or more lines, and is defined by three or more vertexes where those lines meet.

When Euler looked at *polyhedra* (Greek for "many sided" objects), he discovered that the number of corners—vertexes—added to the number of areas—the so-called faces—will always equal the number of edges—lines—plus the number 2.

So let's take our tetrahedron. We know that we have four corners and four areas: $4 + 4 = 8$. We have six edges, to which we add, as Euler tells us, the number 2. That gives us $6 + 2$, or 8: the sum of the corners and faces.

Let's see how this works with a cube, too. But first, I have to say that a cube is like the square in that it has no integrity until triangular braces are added. A cube made of sticks and our flexible connectors simply collapses.

So a cube has eight corners and six areas: $8 + 6 = 14$. If you count the edges, you'll come up with 12, to which we add Euler's number 2, giving us 14, which works out exactly as Euler said.

Now this works for any object you can see. Let's say you take a crocodile and count every exterior point you can see on it, then fill in all the points with lines so that it looks like a crocodile. You'll still find that the number of points plus the number of areas equals the number of lines plus the number 2. It will always hold in every case. Euler has given us a very satisfactory formula in relation to these simple aspects of things you look at. And it even applies to anything you drew as a kid, because anything you can draw has points and lines and areas.

Now in physics they teach you that a cube is the basic unit of volume. But we know by now that a cube cannot exist by itself; it has no stability, no integrity because it is made from squares, which have no stability at all:

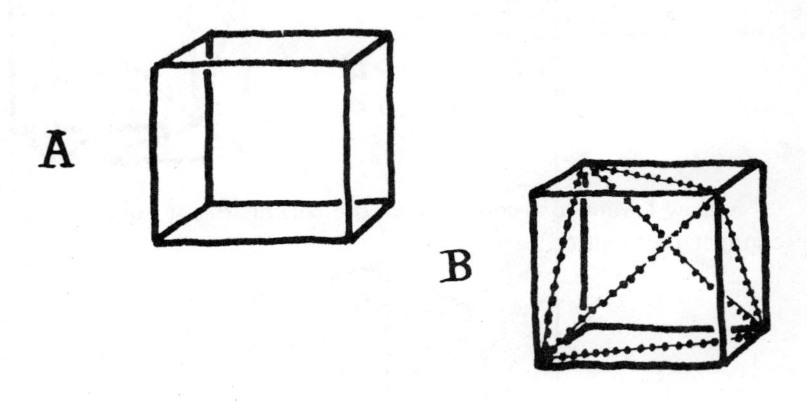

If you want to make a cube (A) hold its shape, you have to put a triangle in each of its faces (B).

So what we have done is put a dotted diagonal on each of the six faces, making each face into two triangles. And if you look closely, you'll see that the shape formed by the six dotted diagonals has four corners, four faces, and six edges. So what makes a cube hold its shape?

JONATHAN: A tetrahedron!

FULLER: Right. A cube won't hold its shape unless it's triangulated. And when you triangulate a cube according to nature's "least possible effort" rule, you get a tetrahedron.

Now if we take our triangulated cube (C), we see that we have connected only four of the cube's eight corners in forming our tetrahedron. That leaves four other, opposite corners. If we connect those unused corners with dash-line diagonals, we find that we get another tetrahedron (D).

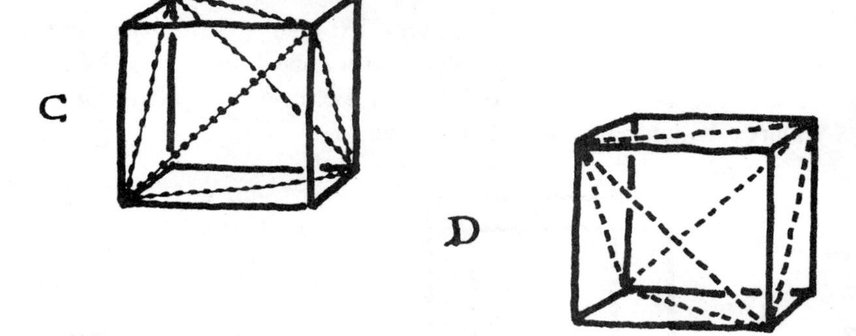

Now I want to account for all the corners of my cube, so I'll connect them all (E):

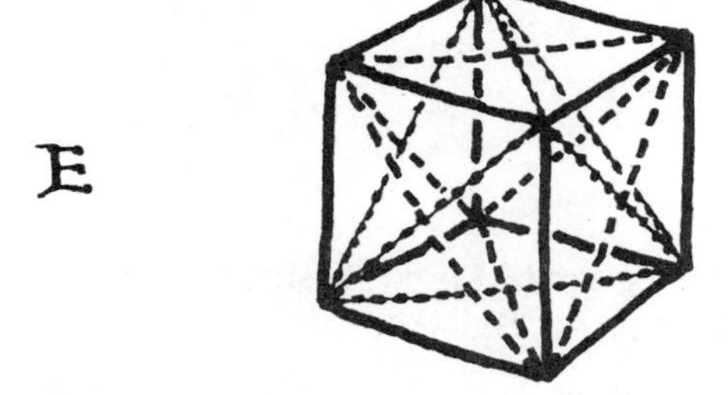

What this gives me is two tetrahedrons, the dotted and the dashed, which we can also call the "positive" and the "negative." What we call the "cube" is really two tetrahedrons, what I call the "star tetrahedron." Let's make one by pulling two of our tetrahedrons together.

RACHEL: Now *that's* what I call a stable cube!

FULLER: It is. In fact, it's the *only* cube.

In my synergetic geometry, I had to give up using the cube, because it doesn't exist. What I do find is the star tetrahedron, made up of the positive and negative tetrahedrons.

Now I'm teaching you nothing that isn't terribly easy to understand right from the moment you start kindergarten. And I've been giving all of you basic physics, even astrophysics. The

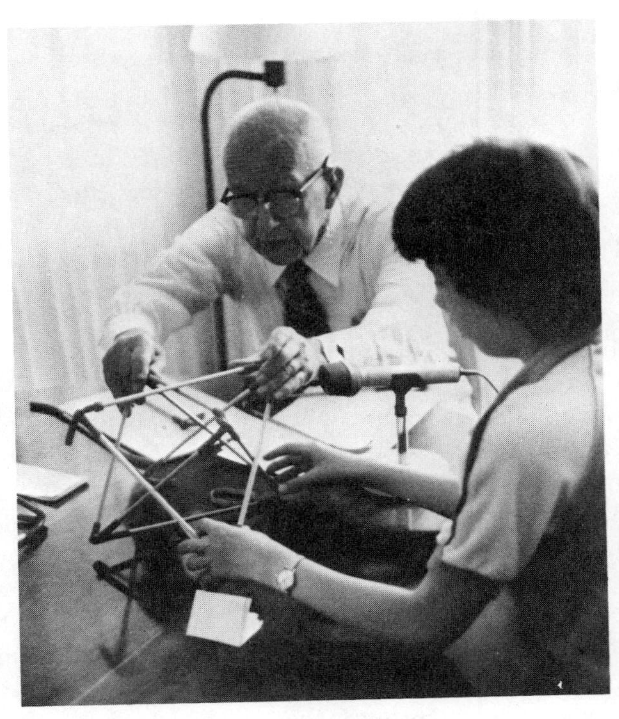

physicists tell you there are no models for basic concepts, only equations. But that's wrong. What I'm showing you here is the way Universe works.

Now the amazing thing about the star tetrahedron is that the two tetrahedrons, the positive and the negative, are always the same dimensions, and each rotates the other like a sphere. And as they rotate, the vertexes—corners—are always describing, outlining, a sphere. Isn't that amazing! It's a sphere of rotation. It's really terribly exciting.

JONATHAN: It's amazing. I can see that I've been taught about something that doesn't really exist.

FULLER: Now let me give you some more of this new way of thinking. And always remember that we are proving our concepts each stage of the way: we are being scientists.

We now have a set of triangle-based structures, each with an inside and an outside, which we call systems. Now a system divides Universe into three parts; all of Universe inside the system, all of Universe outside the system, and the little bit of Universe that composes—makes up—the system that does the dividing.

There's another way of saying this. A system divides Universe into the *microcosm* [the part inside the system], and the *macrocosm,* the rest of Universe outside the system. *Cosm* comes from the Greek *cosmos,* meaning "cosmic" or "universal." *Micro* means "small," and *macro* means "large" or "great." Macrocosm is the greatest extreme outwardness or outsideness of Universe; microcosm is the extreme inwardness or insideness of Universe.

When we're in the realm of nuclear physics, we are dealing with the microcosmic; astronomy deals with the macrocosmic.

So I have a micro and a macro—a non-relevant and a non-considered—for a system is that which is being considered by the mind at this moment out of all the possible considerations in Universe.

Thoughts themselves can be systems. In fact, all of our thinking *is* systemic. I'm having a thought right now. There are things that are too large and too infrequent to what I'm thinking about and there are things that are too small to even be readable. And then there is the system itself, which is what I am thinking about.

This is exactly the way electromagnetic tuning works, like the dial on your radio or the channel changer on your television. There are always bigger waves than you want to tune to, and there are always waves that are smaller. A system is always in between an insideness and an outsideness.

And there's something else to remember about systems. You can't have just *part* of a system without having the whole system. An area has to be the area of something, of some specific system with identifiable, describable characteristics.

Now let's look at something else they taught you in school.

You've been taught to think in terms of "perpendicular" and "parallel," and in "x, y, z coordinates." These are all concepts based on 90°, based on a view of the world where the major relationships are at right angles to each other.

But Universe doesn't operate perpendicular or parallel. Universe operates the way you grow. You grow bigger in *all* directions from the center of your body, like a balloon getting larger and larger as more air is forced in.

So the words for our real Universe are *convergence* and *divergence.* That's how a wave operates.

BENJAMIN: What are the words?

FULLER: Convergence and divergence. To *converge* means to come together. *Verge* is Latin for "moving," and *con* means "together." To *diverge* is to move apart, *di* meaning "apart." That's how all wave phenomena operate. Drop a pebble in a pond, and the waves diverge out from the point where the pebble hit the water. Then the waves hit the edges of the pond and converge back toward the point of impact. Radar operates the same way, and so do sound waves. An echo is sound waves diverging, hitting an obstacle (like a mountain side), and then converging back again so that you hear yourself.

And with us, the way we use our minds, our interests diverge outwardly or converge inwardly. Sometimes we're concerned with world-around problems or interests, sometimes, with some intimate, personal, and private thing. So the way we think is convergent or divergent. We're attracted or repelled, "putting things together" or "tearing things down."

And this is also how we tune things, like a radio, a TV, or a guitar. We tune for exactly the wave we are looking for—the right station, channel, or note—by moving from one side of the wave or the other toward the wave length and frequency we're looking for.

So in our electromagnetic world of wave vibrations, there's no perpendicular or parallel. There's only convergent or divergent. That's the way she operates. So if we are looking for angles, we have to look for the angles of convergence or divergence.

Now let's start with a sphere of a given radius (A), like those spheres generated—formed—by the star tetrahedrons as they rotate around their common center. Now let's take another sphere of the same radius and bring it as close to the first as possible, so that the two are tangent [touching] (B). Now let's take a third sphere and bring it tangent to the first two, so that it is touching both (C). You'll see that the three spheres form a triangular shape, since all three have the same radius.

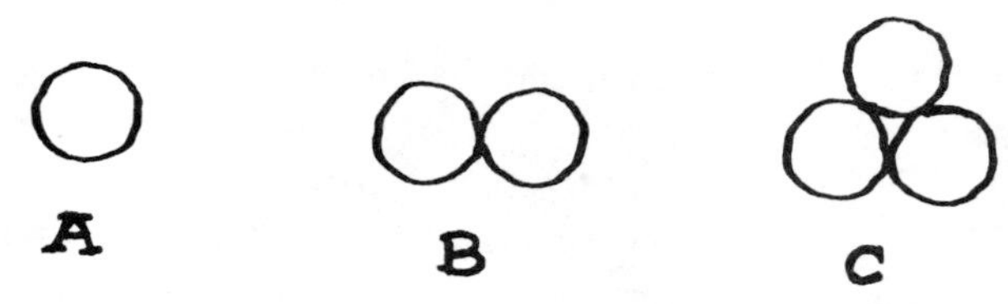

Now if we want to bring in a fourth sphere so that it's tangent with the first three, where does it go? There's only one possible place—on top of the first three, in the little "nest" at the center of the triangular structure. Now when you look at the resulting structure (D), what is it?

JONATHAN: A tetrahedron.

FULLER: Yes. And that's convergence. The spheres have all come together.

Now I've said that you have vertexes, areas, and edges. The balls we have just put together represent the vertexes. The model we make with our sticks and connectors represents the edges. And you can make a model out of paper, just as I'm doing, that represents the areas, the faces.

Now the way these balls pack together is the same way atoms pack together.

Let's look at something else. We'll start with another sphere this time, and we'll see how many spheres of the same size we can place around it so that all the spheres are touching the central one.

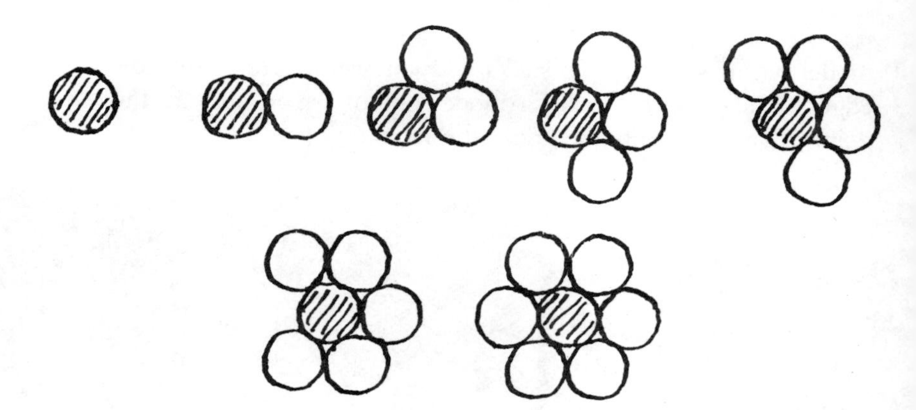

So, with our sphere sitting here on the table, we're able to place six others around it, each touching the central sphere and each touching two other spheres as well, one on either side of it. So I get six-around-one.

Now if we mark the center of each of the spheres and then connect them with lines, we'll see that we get six equilateral triangles forming a pattern described as a *hexagon* (meaning "six-sided").

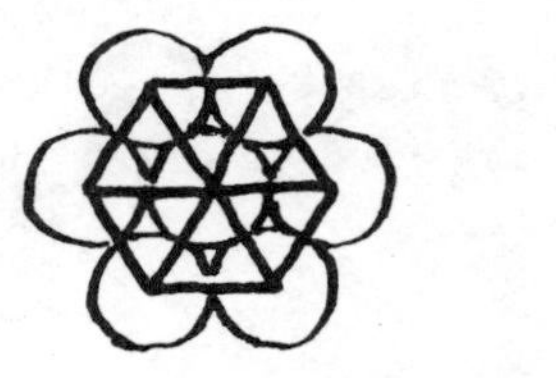

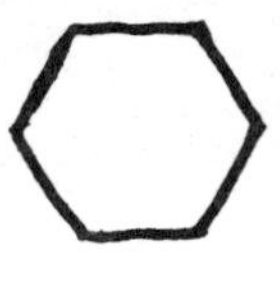

Hexagon

Now the central sphere is what we call a "nucleus," or the central event around which other events converge or diverge. These seven spheres, six-around-one, show us how things converge on the nucleus (A).

But there are still places in our six-around-one where other spheres could sit and still touch the nucleus, our central sphere. Now on our top side there are "nests" just like the nest on the three-ball structure where we placed a fourth sphere to form our tetrahedron.

Now we can place three spheres on top of our six-around-one, all touching other spheres and all touching the central nucleus (B).

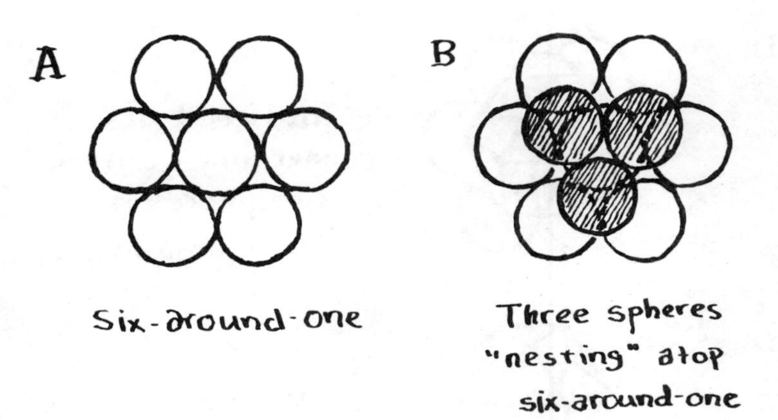

Six-around-one

Three spheres "nesting" atop six-around-one

Now if we turn our structure over, we'll see that there's room for three more on the other side, all touching the nucleus. When we're finished (B), we have a structure in which the central nuclear sphere is completely surrounded by twelve other spheres (C).

Twelve around one

Now I can give you another picture of the same thing by connecting the centers of all the spheres with radius lines, except for the central sphere (D). I call this structure a "vector equilibrium," because all the lines are equal in length, and all the vertexes are equidistant (the same distance) from the central vertex and each adjacent vertex (E).

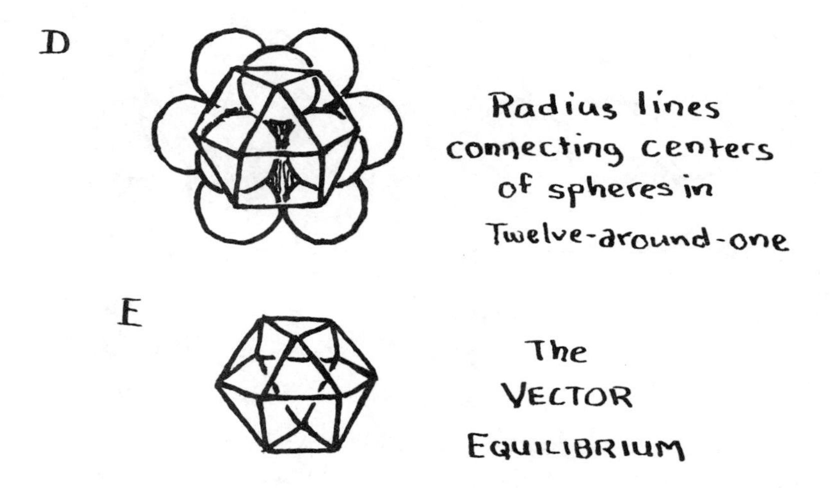

Now let's consider something else. A cube has only three faces.

BENJAMIN: What? It sure looks like six to me!

FULLER: There are three sets of parallel faces in the cube. The cube has only three basic dimensions.

But the tetrahedron has four distinct dimensions, four distinct planes. So it is four-dimensional. And the dimensions of the tetrahedron are based on the real-life example of closest-packed spheres.

The tetrahedron's four dimensions are based on 60-degreeness, rather than the 90-degreeness of the cube. And dimensionality is based on 60-degreeness, not on 90-degreeness.

Now let's look at our tetrahedron again. Each of its faces is one of the four planes of 60-degreeness, and directly opposite each face is a vertex. Now let's think about moving one of these planes or faces toward the opposite vertex. I'll shade the plane so you can see it (A). Now as we move closer and closer to the vertex, the plane gets smaller and smaller (B, C).

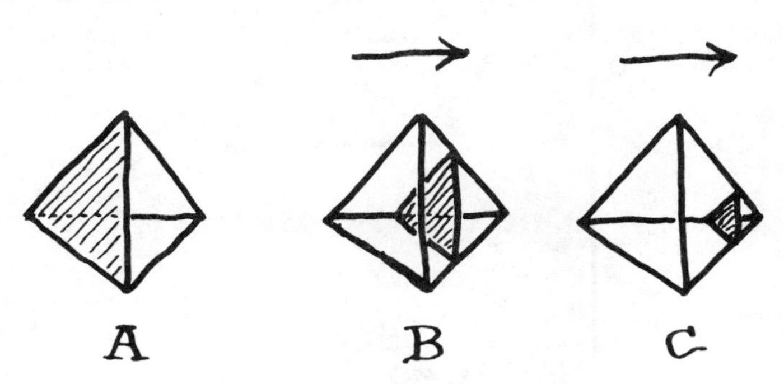

Now if we move all four planes to the center at the same time, we'll have four planes passing through the same place at 60° to each other (D). That's exactly what's happening with the vector equilibrium.

If we go back to the drawing [on page 70] we can see that the six-around-one from which we began our vector equilibrium forms a shape we call the hexagon. And if we look at drawing E [on page 74] we can see that the completed vector equilibrium outline is composed of four hexagons interlocked with each other. Each of these hexagons is at 60° to the other, and the plane of each hexagon passes exactly through the nucleus.

One of the six
hexagonal planes
of the
vector equilibrium

So the four hexagonal planes of the vector equilibrium are identical to the four 60 ° planes of the tetrahedron when they are all moved to the center of the structure.

In the vector equilibrium there are four hexagons crossing each other at the same time in 60 ° relation to each other. Can you all see that?

ALL: Yes.

FULLER: So the vector equilibrium has four planes all passing through one central nucleus. It has twelve corners, and it's based on the closest packing of same-sized spheres, twelve-around-one.

This is the way atoms pack. This is the most basic part of the atomic nucleus. This is the way real things converge and diverge.

Now here's a way we can make a model of the planes of the vector equilibrium from four same-sized paper circles and twelve bobby pins. Measure your circles on four sheets of paper so they are exactly the same size. Use a compass—or if you don't have a compass, then use the bottom of a large can or the rim of a bowl.

Now, with your four paper circles, follow these steps:

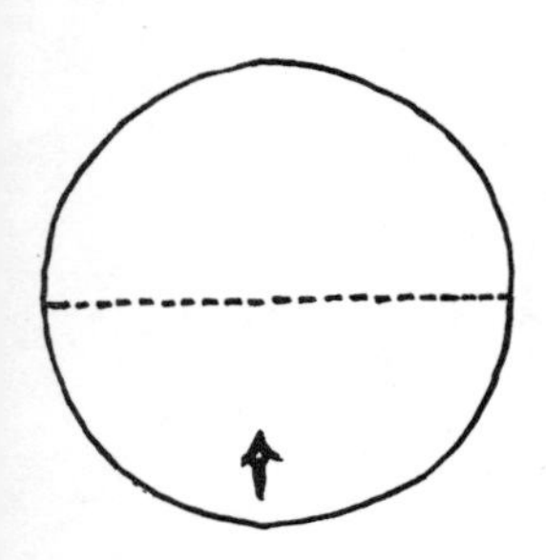

Fold lower half over upper . . .

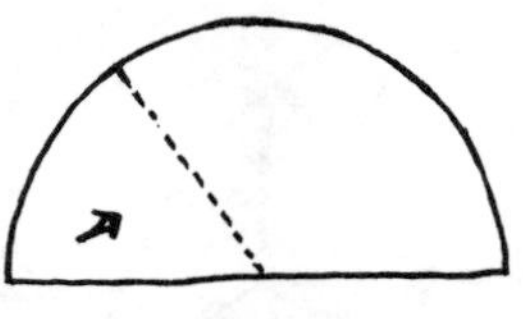

. . . then fold at one-third, bringing folded section toward you . . .

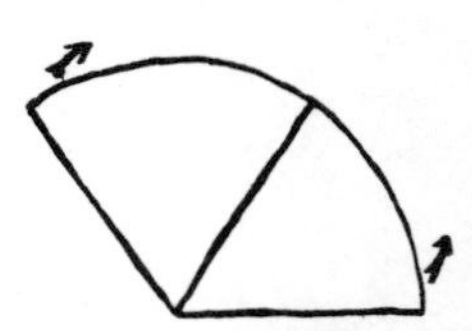

. . . folding again where edge divides remaining area in half, folding both sections away from you . . .

. . . resulting in a three-layer triangular sandwich" of doubled-over triangular segments . . .

. . . that looks like this "Z" shape when viewed from the curved edges . . .

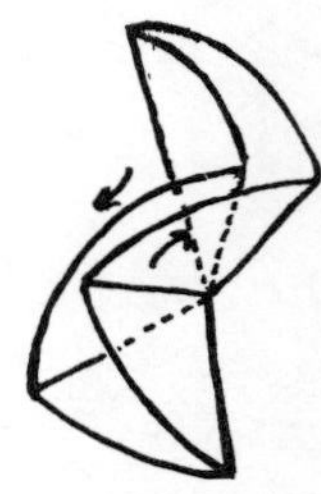

and that unfolds . . .

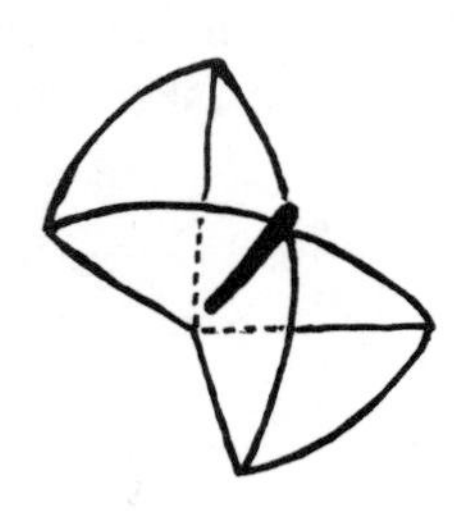

. . . to form a BOW-TIE, which is stabilized by a bobby pin.

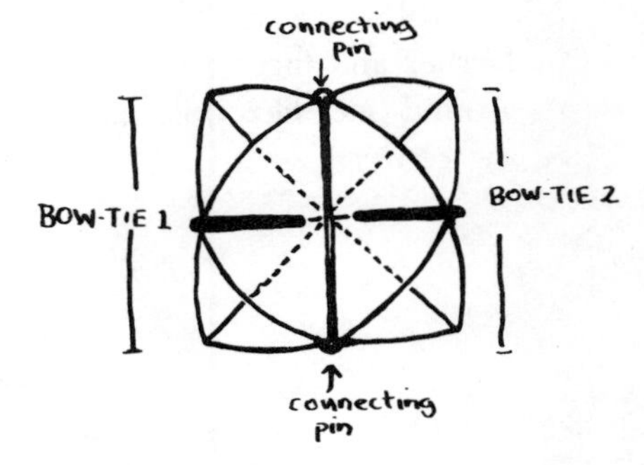

Two BOW-TIES joined by two more bobby pins

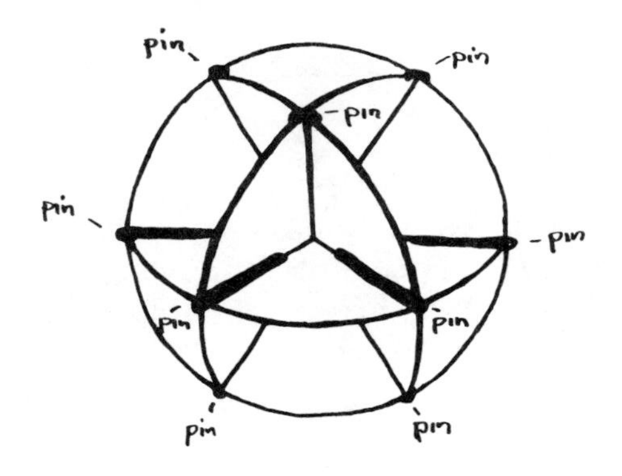

Two Two-BOW-TIE Units joined at the four here-to-fore unpinned triangular-point "feet" become the planes of the

VECTOR EQUILIBRIUM

Now if you'll look at the drawing D on page 73, you'll see these same planes—only we formed them from a tetrahedron.

Now we can keep expanding our vector equilibrium model farther and farther out. We can add layers to our basic twelve-around-one, like this, looking at a cross-section (slice) through the center:

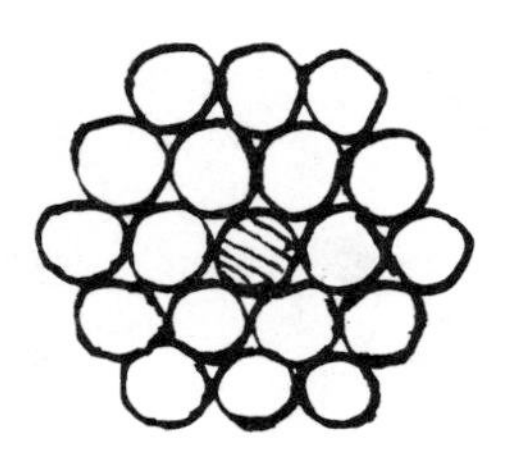

We find that to completely cover our first ball it took twelve new balls. And to form a layer that completely surrounds the twelve balls of the twelve-around-one takes forty-two new balls, nested on the first twelve just the way we nested the fourth ball to form the tetrahedron. Now a layer to completely cover the forty-two-ball layer will require ninety-two new balls. The next layer would be 162, the next 252, and so forth.

Now if we go back and look at drawing E on page 74, we can see that the surface of our vector equilibrium is formed of two shapes—triangles and squares. If we count, we'll see that there are eight triangles and six squares.

Each time we add a new layer of spheres to our vector equilibrium, it maintains the same pattern of triangles and squares.

Now we've discovered that each time we add a layer of spheres around a central nuclear sphere, the number of spheres in every layer ends with the number 2.

1 layer	=	12 spheres
2 layers	=	42 spheres
3 layers	=	92 spheres
4 layers	=	162 spheres
5 layers	=	252 spheres
6 layers	=	362 spheres

Now remember when we talked about Euler's formula [pages 61–62]? Well, Euler's formula laid some importance on the number 2. The number of the vertexes plus the number of the faces equals the number of the edges plus the number 2.

Now the number of spheres on the surface of a vector equilibrium is the number of the vertexes, and we have already established that the center of one of our spheres represents a vertex.

When we look at the number of spheres required to make each new layer of spheres on our vector equilibrium, we discover a new formula that is a corollary of Euler's formula.

First, let's subtract the number 2 from the number of spheres in each layer:

$$
\begin{array}{lrcl}
\text{1 layer} & 12 - 2 & = & 10 \\
\text{2 layers} & 42 - 2 & = & 40 \\
\text{3 layers} & 92 - 2 & = & 90 \\
\text{4 layers} & 162 - 2 & = & 160 \\
\text{5 layers} & 252 - 2 & = & 250 \\
\text{6 layers} & 362 - 2 & = & 360
\end{array}
$$

Now I suspect you can begin to see a relationship between these numbers and the number of layers out from the center. That relationship will become a lot clearer if we take the additional step of dividing each of these new numbers by 10.

$$
\begin{array}{lrcl}
\text{1 layer} & 10 \div 10 & = & 1 \\
\text{2 layers} & 40 \div 10 & = & 4 \\
\text{3 layers} & 90 \div 10 & = & 9 \\
\text{4 layers} & 160 \div 10 & = & 16 \\
\text{5 layers} & 250 \div 10 & = & 25 \\
\text{6 layers} & 360 \div 10 & = & 36
\end{array}
$$

Now it becomes clear. If we raise 1 to the second power (1 times itself, 1×1) we get 1; 2 to the second power is 4; 3 to the second power is 9; 4 to the second power is 16; 5 to the second power is 25; 6 to the second power is 36.

Now before I state our new formula, let me give you one more term: *frequency.* When I speak of frequency in the vector equilibrium, I am referring to the number of layers of spheres out—diverging—from the central nuclear sphere. So our first layer of 12 spheres is the 1-frequency; our second layer of 42 is the 2-frequency; the third layer of 92 is the 3-frequency; and so forth.

And now on to our new formula.

I find that the number of spheres in any frequency of the vector equilibrium is equal to the number of the frequency (F) squared multiplied by 10, plus the number 2. We can write it out this way: $N = 10F^2 + 2$, where (N) is the number of balls in the layer and (F) is the number of the frequency.

Let's see how this works. If we want to find the number of balls in the 4-frequency layer, then we square the number of the frequency—$4 \times 4 = 16$—multiply that number by 10—$16 \times 10 = 160$—and then add the number 2—$160 + 2 = 162$. This works for every frequency of the vector equilibrium.

What we're discovering here is very exciting, because it relates to things going on in the real world and not just imaginary concepts.

Now let's take a special look at the 3-frequency vector equilibrium. There are 92 spheres in the outer layer. You may have already learned in school that there are 92 naturally occurring chemical elements in the known universe. And if we add up the numbers of the first three frequencies (12, 42, and 92) we get 146—and that works out to be the number of neutrons in uranium, which is the 92nd, or final, chemical element. So I find we're already dealing here in nuclear physics, very deeply into it, further than most of the physicists have gone themselves. There's also something else about this layer: it's the first in which the arrangement of the spheres makes the central sphere totally invisible. (You can still see glimpses of it through the spaces in the 2-frequency.)

We are dealing here with concepts and principles, and these are independent of size. The concept "triangle" has no particular size. A triangle is a triangle, whatever its size may be.

Now I said a system divides the universe into an inside and an outside. It is triangular, and a tetrahedron is the minimal structural system or "something" in the universe. There is another triangular system that is composed of all triangular faces, which divides the universe into an inside and an outside, and which exactly fits up against the tetrahedron. This new system is called the "octahedron."

Octa means "eight," so octahedron means eight-sided. And our octahedron has eight faces, each an equilateral triangle. There are six vertexes, and twelve edges. Going back to Euler's formula (corners + faces = edges + 2) we have 6 corners + 8 faces = 12 edges + 2, or 14 = 14.

Let's take a look at this new system.

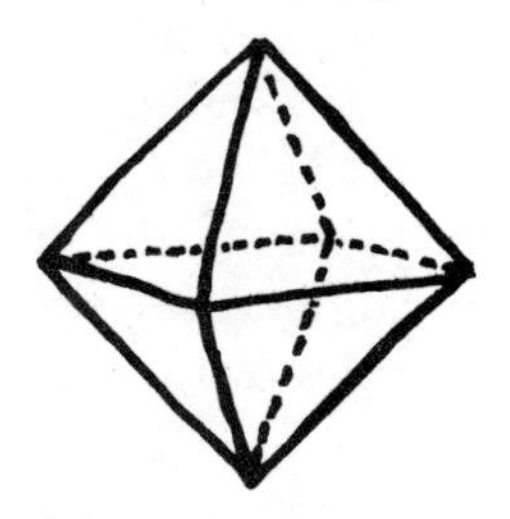

THE OCTAHEDRON

There's something else about the octahedron and its relationship to the tetrahedron. These two systems in combination with each other work to fill all space, leaving no gaps. They are what I call "allspace filling."

I call this combination of the octahedron and tetrahedron the "octet truss."

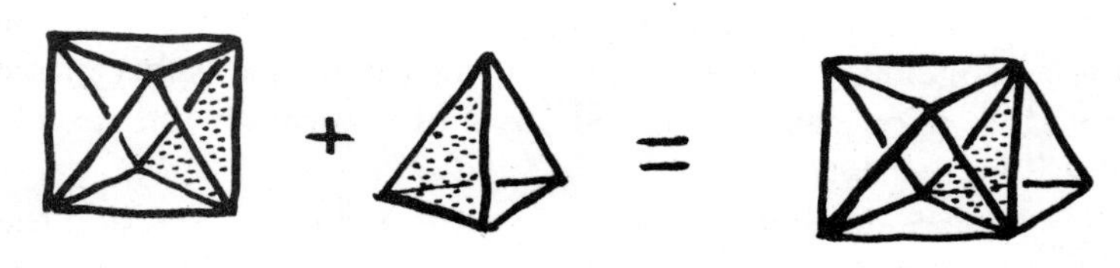

OCTAHEDRON + TETRAHEDRON = OCTET TRUSS

The Greeks liked cubes because they seemed to fill all space, packed next to each other cube to cube.

But we've already seen that the cube is not stable. It's not triangulated, and so it's not able to hold its own shape. But the octet truss is based on 60 ° triangulation and is stable.

Now if you double the size of a cube, you increase the volume eight times. That's easy to see in this drawing:

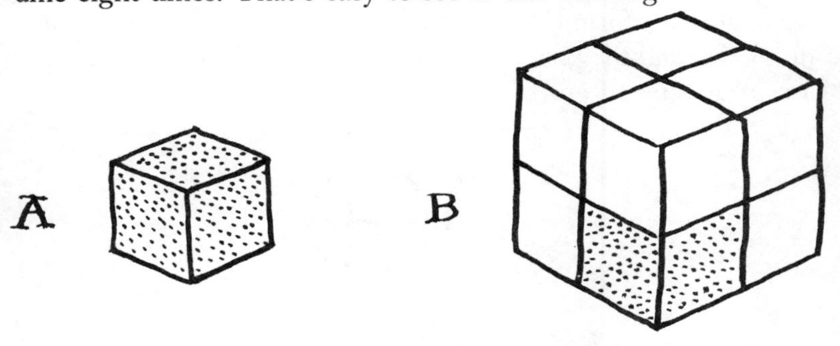

If our cube is one inch on a side, we double it by making a new cube two inches on a side. Our first cube (A) had a volume of one cubic inch; our doubled cube (B) has a volume of eight cubic inches. [See also pages 123–125.]

Now let's see what happens when we double the size of a tetrahedron (C) to (D).

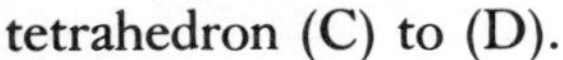

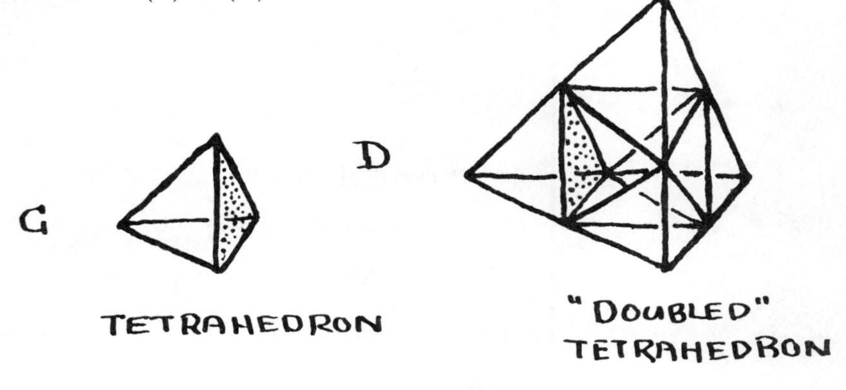

Our new doubled tetrahedron also has a volume of eight. It's a rule in geometry that when you double (square) the edge dimensions of a solid, you "cube" (increase eightfold) the volume.

Now something interesting happens with our tetrahedron when its edge dimensions are doubled. If you divide each of the edges of the doubled tetrahedron in half and connect the midpoints, you get a structure in which each of the four corners is composed of a tetrahedron the size of the tetrahedron (C) which we doubled to form (D). If we remove these four corner tetrahedrons, we find they have been covering four of the eight faces of a central octahedron (E):

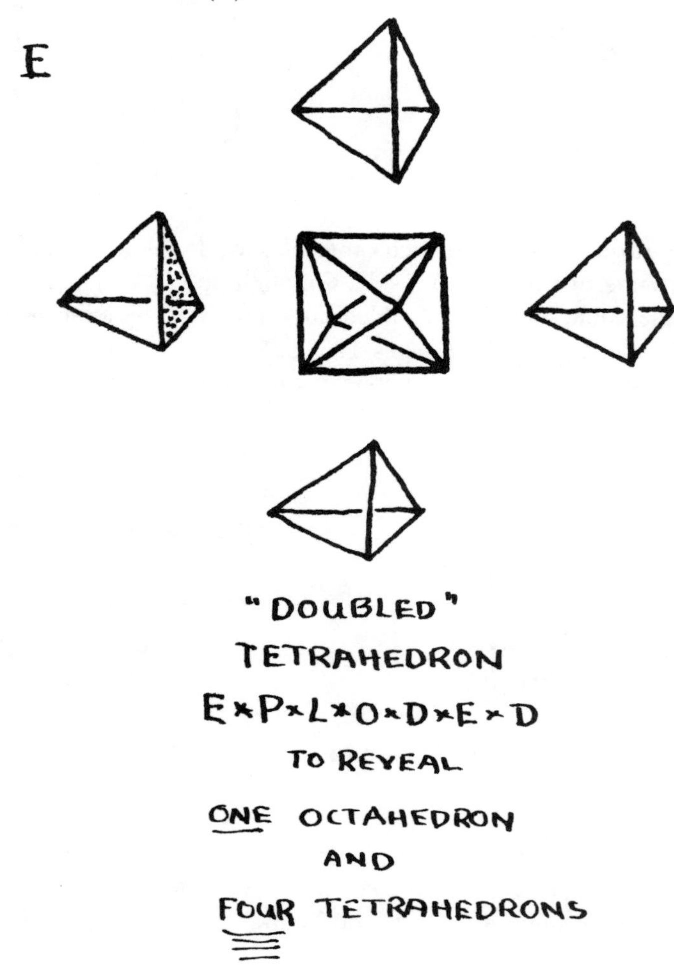

Since the volume of each of the four small tetrahedrons is the same as that of our starting tetrahedron (C), we know that when we remove these four from the doubled tetrahedron we are removing four units of volume. We've already established that the volume of the doubled tetrahedron is eight times that of the original tetrahedron, so eight units minus four units (the four tetrahedrons) leaves four units—which is the volume of the octahedron. So the volume of an octahedron is four times that of a tetrahedron with the same dimension on any given edge.

Now the octahedron has six corners (vertexes), each vertex being directly opposite another. This gives us the x, y, z coordinates, or what is called "square symmetry." This is where the square *does* come in in nature, but only because it is part of the structure of the triangulated octahedron.

The octahedron has three planes, each at 90 ° to the other, each shaped like a square. All the angles in the same edge [circumferential] plane are 90 °—but each face is 60 ° from any adjacent, tangential [touching] face.

If we draw lines connecting the opposing vertexes of the octahedron, we get three axes at 90 ° to each other [heavy lines].

Now I can divide the octahedron into identical one-eighth units, using the three central 90 ° axes we have just seen. If we slice the octahedron along its square circumferential—edge outline—planes, it divides the octahedron into eight parts, each having one of the original 60 ° triangular faces, and three inner angles of 90 ° each.

Three 90°
planes to cut the
OCTAHEDRON
into 1/8th sections

Now what we have here is the corner of a cube. The three 90 ° angles which came from the center of the octahedron now become the outer corner of a cube.

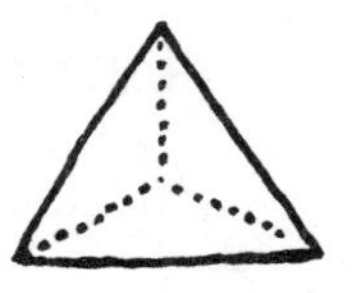

1/8th OCTAHEDRON

Since the volume of our octahedron is eight (compared with one for the starting tetrahedron), the volume of a one-eighth octahedron would be one-half.

Now if I take four of these one-eighth octahedra and attach them to a tetrahedron so that the triangular faces are exactly touching and the 90°-90°-90° corners are pointing outward, I get a cube.

Now let's look at what we did. We took four one-eighth octahedra, each with a volume of one-half, and added them to a tetrahedron, with a volume of one. That gives us a volume of $4 \times \frac{1}{2} + 1$, or a total of 3. So the volume of a triangulated cube in which the diagonal of any face is the same length as a tetrahedron is three times that of the tetrahedron. So a cube has a volume of three.

At this point, let's look at one more thing about the octahedron. If you look directly at one of the triangular faces so that it is directly facing you, and look through the face to see the face opposite, you'll see that the opposite face is a triangle pointed in the opposite direction from the near face, the one you are looking through. So the two faces will form a shape like a six-pointed star. In our drawing the opposite, farther face is shaded. I want you to bear this is mind, because it will help you identify the octahedron in some of the drawings I'm about to make to help explain some more new ideas to you.

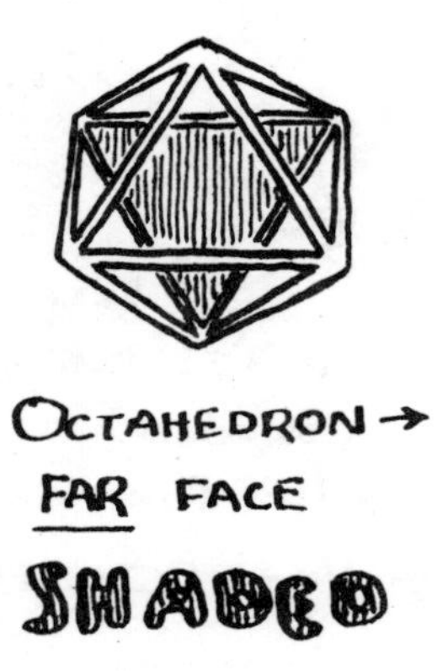

We have already discovered that when we doubled the edge length of a tetrahedron and connected the midpoints of the edges, we formed a new tetrahedron with a volume eight times that of the original and composed of four tetrahedrons the same size as the original and one octahedron. We also discovered that the volume of the octahedron was four times that of the tetrahedron of the same face/edge size.

Now if we make another tetrahedron with an edge length three times that of the original and divide the edges into thirds and connect the dividing points, we'll get a large tetrahedron which has an octet truss internal structure composed of four octahedrons and eleven tetrahedrons. So the volume of our tripled tetrahedron would be 16 (four octahedrons each with a volume of four [$4 \times 4 = 16$]) plus 11 (eleven tetrahedrons, each with a volume of one [$11 \times 1 = 11$]), for a total of 27, which is 3^3 ($3 \times 3 \times 3$, or 3 "cubed").

Now let's take a look at some one-layer octet trusses and see what we can discover about them.

The basic model for our octet truss is one octahedron and three tetrahedrons, which looks like this from above:

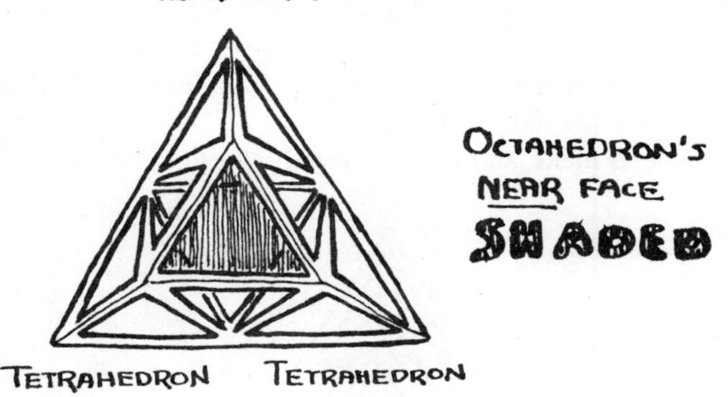

To help our accounting, I'm shading the near triangular face of each of the octahedrons, so we can readily identify them.

So here we have three tetrahedrons and one octahedron. This gives us a volume of 3 + 4, or 7.

Our basic truss had an edge length of two. Now let's make the next larger one-layer truss, which has an edge length of three.

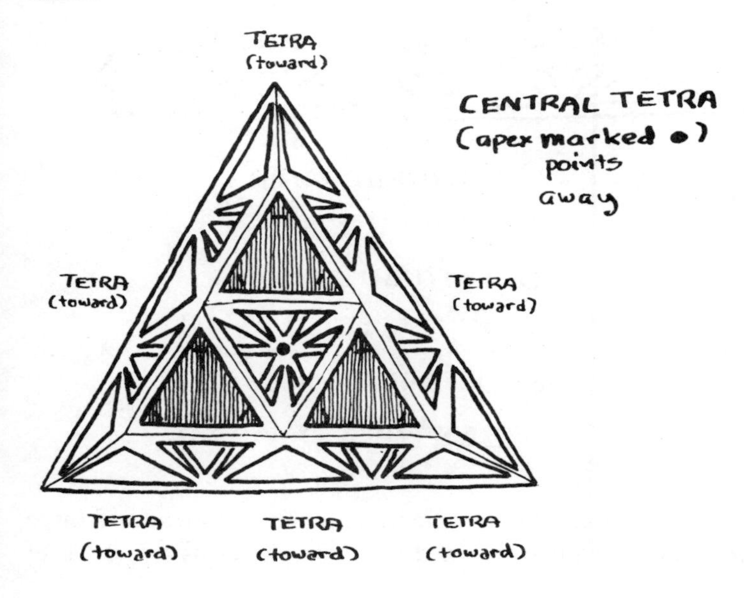

Here we have six tetrahedrons pointing toward us, one pointing away, and three octahedrons. This gives us a volume of seven—our tetrahedrons—plus twelve—our three four-unit octahedrons—for a total of nineteen.

Now let's make another truss with an edge length of four.

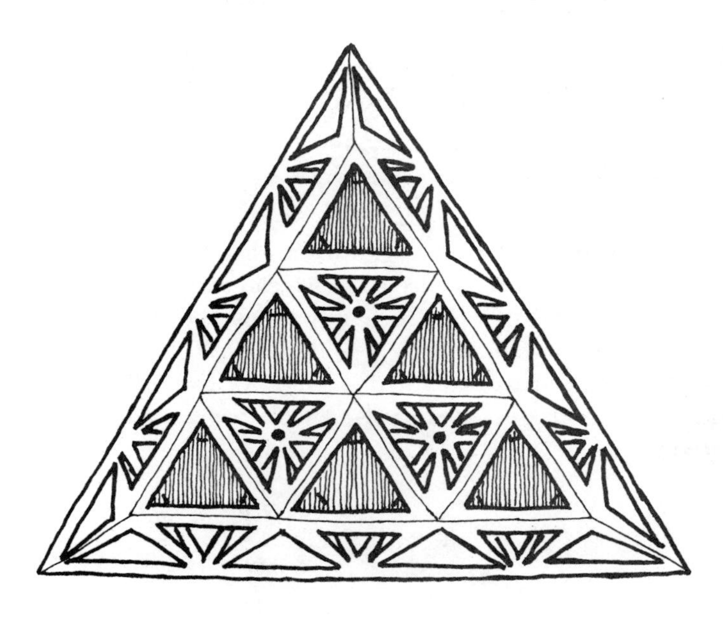

Here we have ten tetrahedrons pointing toward us, three pointing away, and six octahedrons. This gives us a volume of

10 + 3 (the tetrahedrons) plus six four-unit octahedrons, for a total of 37.

So now I can set our 19-volume truss and set it on top of the 37-volume truss we've just made. Then on top of the 19-volume truss I'll place the 7-volume truss. Last of all, we'll place a 1-volume tetrahedron on the very top. This give us our four-module tetrahedron, a tetrahedron with an edge length four times that of the original. The volume of a four-module tetrahedron is 37 + 19 + 7 + 1, or 64.

The 4-MODULE (3^4)
TETRAHEDRON
(apexview)

Volume = 64 1-MODULE
TETRAHEDRONS

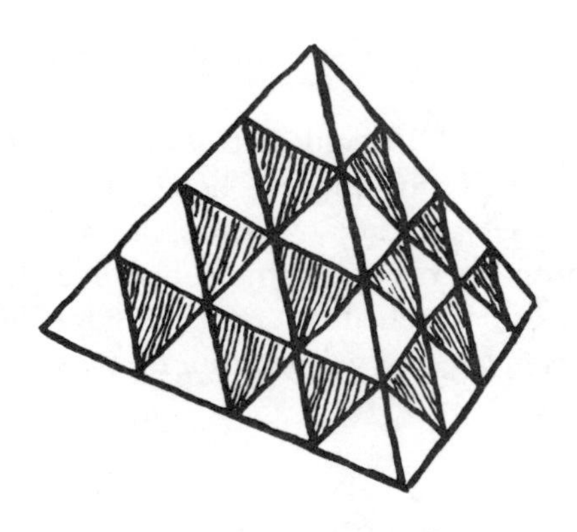

Simplified view
4- MODULE (3^4)
TETRAHEDRON
(edge view)
exposed octa faces
shaded
tetra faces solid
white

Let's look at what we've done in another way.
We started with 1.
Our next level is 7, so $1 + 7 = 8$, which is 2^3.
Next came 19, and $19 + 8 = 27$, which is 3^3.
Last came 37, and $37 + 27 = 64$, which is 4^3.

So our basic tetrahedron is 1, and 1^3 is still 1. Next came $7 + 1$, or 8, which is 2 to the third power (2^3). Then came $19 + 8$, or 27, which is 3 to the third power (3^3). And then came $37 + 27$, or 64, which is 4 to the third power (4^3).

This gives us a mathematical law about our tetrahedrons: We can say that the volume of a tetrahedron equals the module length to the third power.

So instead of saying "cubing" when we talk about raising a number to its third power—multiplying a number n × n × n—we can say "tetrahedroning," because these are volumes expressed in tetrahedrons, and they coincide exactly with the third powers of the edge modules.

Because nature is always most economical, and because a cube uses three times as much Universe as necessary (a cube has a volume of three compared with a tetrahedron of the same diagonal module), and because a tetrahedron can hold its shape while a cube cannot, it is obvious that nature is tetrahedroning instead of cubing.

Because physicists started out with the imaginary, unstable cube as their model instead of the real-world stable tetrahedron, they got into all these imaginary numbers and other complicated and completely unnecessary mathematics. It would be so much simpler if they started out with the tetrahedron, which is nature's best structure, the simplest structural system in Universe.

(Just as an aside, to remember later when you're studying physics in school, I want to point out that the tetrahedron is also equivalent to the quantum unit of physics, and to the electron.)

Now let's talk about one last system.

In our tetrahedron, we find three equilateral triangles meeting at every vertex. The sum of the angles meeting at each vertex is 180° (60° + 60° + 60°). Now in the octahedron we have four equilateral triangles meeting at each vertex; and the sum of the angles meeting at each vertex is 240° (60° + 60° + 60° + 60°). Now there's one more system, and it has five equilateral triangles meeting at each vertex, with a sum of angles for each vertex of 300° (60° + 60° + 60° + 60° + 60°).

JONATHAN: Could you have one with six triangles?

FULLER: No, because that would be 360°, which would be a plane going out to infinity and not coming back on itself to divide Universe into an inside and an outside.

So there are only three structural systems in Universe: tetrahedron, octahedron, and this new one, which we call the "icosahedron" (*icosa* means "twenty" in Greek). All crystallography (the science of how atoms and molecules arrange themselves into regular patterns) comes back to this. These are the only ways in which atoms interact.

I find it most interesting that they don't teach you about this in school at all.

Now the icosahedron looks like this:

ICOSAHEDRON

There are five triangles around the top, ten around the Equator, and five around the bottom.

And if you look at it directly from any vertex, it looks like this:

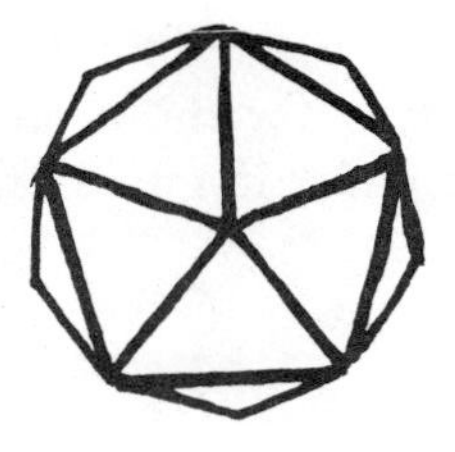

FIVE TRIANGLES meet at every vertex of the ICOSAHEDRON

Now the tetrahedron has six edges; the octahedron has six-teen; and the icosahedron has thirty. Each of these numbers can be divided by 6:

Tetrahedron	$6 \div 6 = 1$
Octahedron	$12 \div 6 = 2$
Icosahedron	$30 \div 6 = 5$

So I say that you can't have a system unless the edges are six or a multiple of six. In any given units of energy, there are six edges. These represent the push-pull forces or vectors that are necessary for all structures.

Six is the beginning of a structural system. We have one set of six in the tetrahedron, two sets in the octahedron, and five in the icosahedron. In the tetrahedron, one set of six gives a volume of one; in the octahedron, two sets of six give a volume of four; and in the icosahedron, five sets of six give a volume of approximately twenty. So with one set of six, you get a ratio of sets to volume of 1:1; with two sets, the ratio is 2:1; and with five sets, the ratio is approximately 4:1. So the icosahedron gives you the most volume enclosed for the least material used—or for the "least structural investment," to use engineering terms.

It is the icosahedron, this most economical of nature's structures, that I have used as the basis for my geodesic domes. I might also note that we have also found nature to be using the icosahedron as the structure for the protein shells that contain the RNA/DNA genetic codes in our cells.

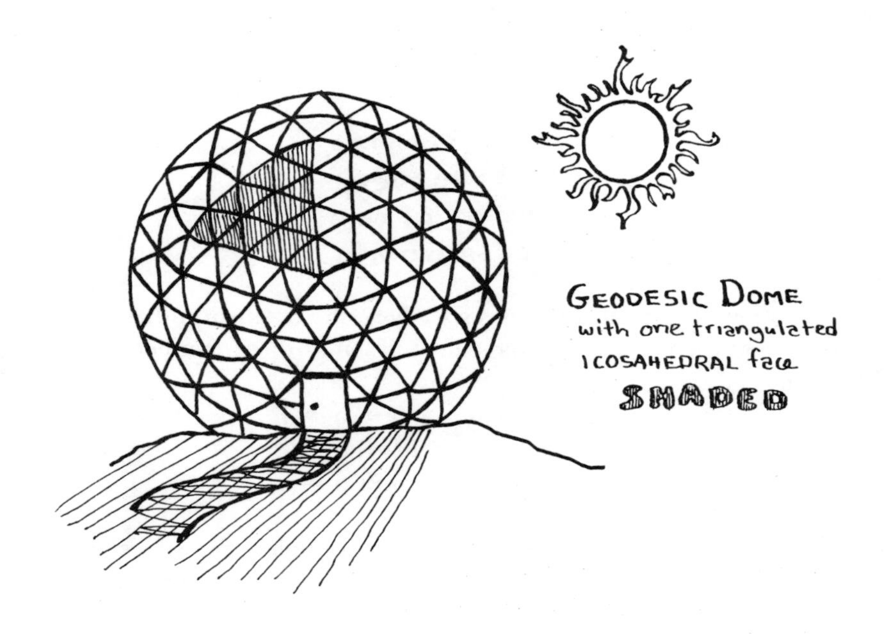

Now all of you are used to the idea of building by putting bricks on top of bricks. When you are very little, they give you wooden building blocks, and that's how you use them. Now this way of building I call putting compression on top of compression. And when I looked at nature, I saw she wasn't building this way.

When we see a brick, it looks solid to us. But it's not. It's really full of all kinds of atomic arrangements, in patterns like those we've been looking at.

And so I asked myself, "How is nature *really* doing things?" And I could see that the moon goes around the earth but never touches it, just as the earth is going around the sun but never touching it. Yet all of them are held precisely in place as surely as if by bricks and girders. I saw that nature is using *tension*—you can call it gravity—to hold things in place. Nature has islands of compression—like the earth, or an atom—held together by invisible but continuous tension.

Compressions are discontinuous—they don't touch each other—and are held in place by tension, which *is* continuous.

Now for an example. A long time ago, wheels were completely solid, made of thick slabs of wood bolted together. These wheels were *compressions.* But when we get to wheels with wire spokes, we made a *tensional integrity.* We have the rim, a band of compression, and the hub, an island of compression, held together in tension by the spokes, and that's more or less nature's way of doing things.

CART WHEEL

WIRE WHEEL

Now we can also make this same sort of structure omnidirectionally, as we'll do right now.

If you look at our structure, you'll see twelve five-sided (pentagonal) shapes. They correspond exactly to the vertexes in our old friend, the twelve-around-one vector equilibrium.

We have these sticks, these compression members, with Dacron threads between them (we use Dacron because it doesn't

stretch like other fabrics do). No thread is loose. These pentagons are everywhere the same. The tension is equally distributed. It is like a pneumatic—inflatable—tire or ball where the atmospheric molecules inside are hitting the tension network and not being able to escape.

So our structure is held together by tension, and this is the way Universe is put together.

That is also why I made my geodesic structures based on the icosahedron—to use the least material for the most volume enclosed—and that in turn gives the geodesic structure the purest possible tensional integrity. Now I've given this concept a shorter name—*tensegrity*, from *tens*ional int*egrity*.

Now I think that's enough for me to establish with you how and why I've given you a way of reassessing what it is we *really* were learning, to make what you're learning coincides with experience and then to turn it into an advantage of ours [just like the advantage of the lever] so we can get more housing for humanity—more environmental control—for less materials. And with this system, I can do it.

I can turn what we are learning here to great advantage. We need to rehouse humanity, and we can do it with new structures based on these principles. There's a limit to cross-section in compressional structures, to the width of structures that can be built with conventional methods. The height of a steel column can't go over forty times its diameter before it starts to curve over like a banana. But there's no limit of tension to cross-section. To make a longer and longer suspension bridge, you simply need a better alloy.

When I get into tensegrities, there's no limit to the clear span that can be enclosed. [A clear span is the area enclosed or covered by a structure without internal supporting columns.] When I started geodesics, the largest clear-span dome in the world was only 150 feet in diameter. Today we could make one a half-mile in diameter. In fact, we could make one to go around the world if we wanted. There's *no limit.*

Now there's one last thing I'd like to show you.

I wanted to be able to see the world correctly, but I found that there were problems with most maps. For instance, the one they give you most commonly is called a "Mercator projection."

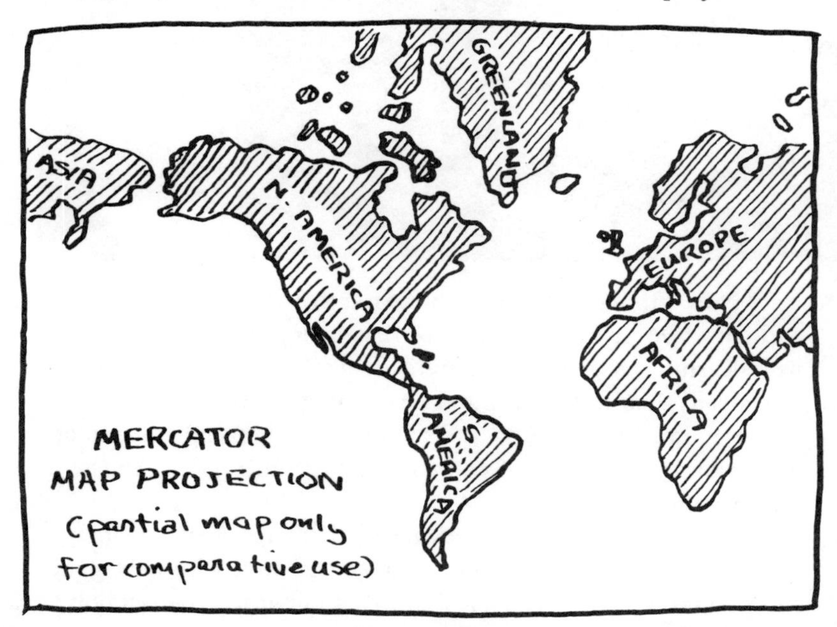

Now if you look at a Mercator map, you'll find that Greenland measures twice the size of South America, and North

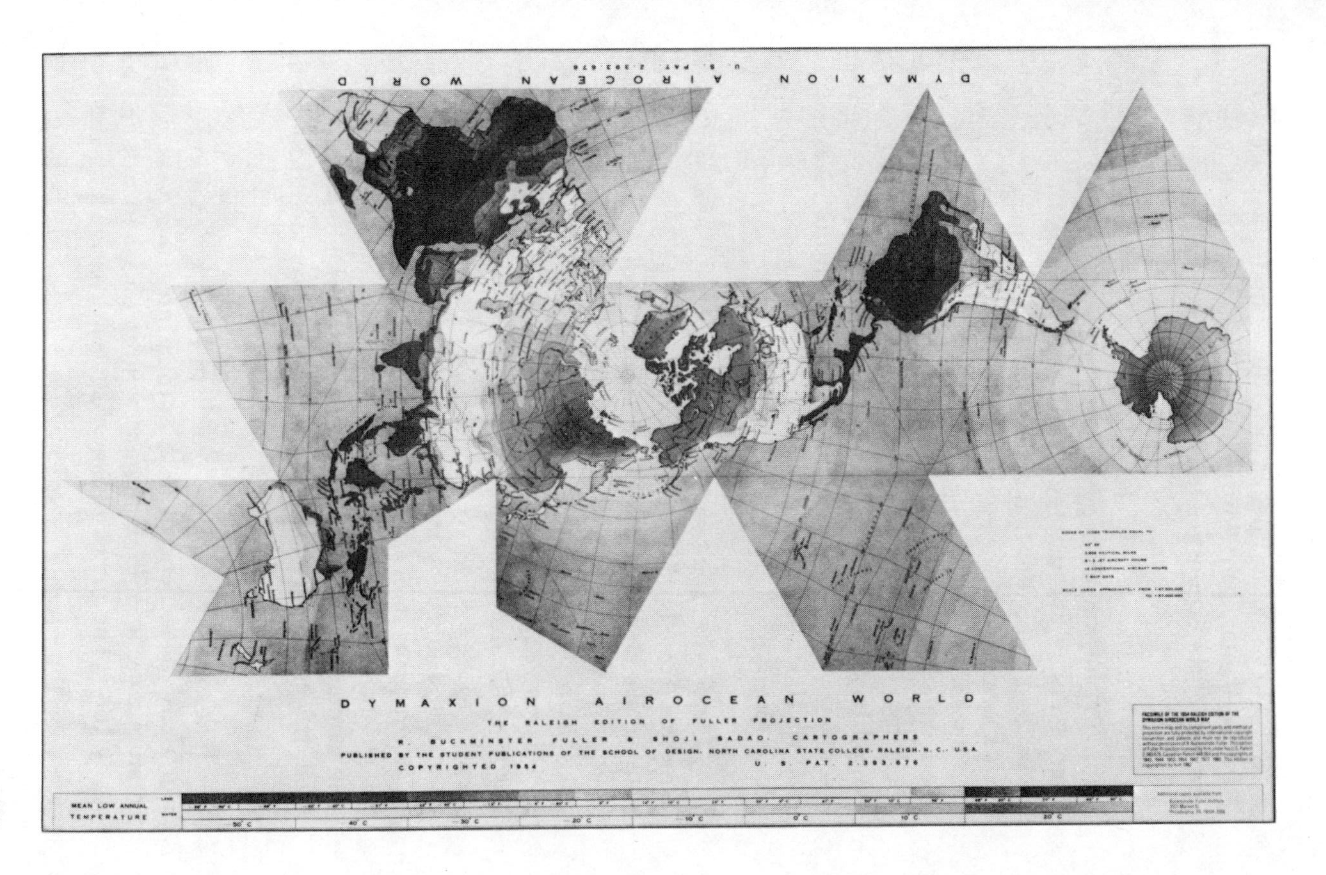

DYMAXION AIROCEAN WORLD
U. S. PAT. 2,393,676
DYMAXION AIROCEAN WORLD
THE RALEIGH EDITION OF FULLER PROJECTION
R. BUCKMINSTER FULLER & SHOJI SADAO, CARTOGRAPHERS
PUBLISHED BY THE STUDENT PUBLICATIONS OF THE SCHOOL OF DESIGN, NORTH CAROLINA STATE COLLEGE, RALEIGH, N.C., USA
COPYRIGHTED 1954
U. S. PAT. 2,393,676
MEAN LOW ANNUAL TEMPERATURE
LAND
WATER
50° C
40° C
30° C
20° C
10° C
0° C
10° C
20° C

America looks bigger than Africa. But both of these are wrong. Africa is bigger than North America, and Greenland is really a fraction of the size of South America.

When I looked at other projections, they had similar problems. In one way or another, they provided distorted information. I discovered that most maps are very false, misleading maps. So I wanted to find a new map, a better map.

I began with the twenty triangles of the icosahedron superimposed on the surface of our spherical earth. Now when these triangles are spherical, each angle is 72 °. When I reduced them to a plane, each angle became 60 °. So to correct for any possible distortions, I simply hold a uniform boundary scale when I contract all the corners symmetrically. What this gives me is a map which lets me see the whole world at once without any visible distortions.

Here's the *right* size of Greenland. The right size of South America. Here is *all* of Antarctica, which you can't even see on most maps. Here is a map with no breaks in the continental contours. I have one world-island in one world-ocean, without any break in the land. It's the first time you've been able to see all the world *as it really is* at once, without any visible distortion of the shape or the size of the parts.

The shadings on this particular map I'm showing you now relate to temperature. Now water holds its temperature much longer than does crystalline material—soils and rocks—so the temperatures over the water are much more uniform than over land. It gets much colder and much hotter over land than water.

The "cold pole" of the Northern Hemisphere is in Siberia near a city called Verkhoyansk. In the winter, the average temperature there is −58 ° Fahrenheit. But in the summer it gets to nearly 100 °. The annual variation between average high and low

temperatures is almost 150 °. But when you get down around the Equator, the variation may be as little as 20 °. Now the temperature on the average August day at Verkhoyansk may be no different than the average temperature that same day on the Equator. The real extreme variations have to do with cold, not heat. The real difference between places is how cold they get, not how hot. And this map is shaded to show the mean low annual temperatures.

There are other things the Dymaxion map can tell us.

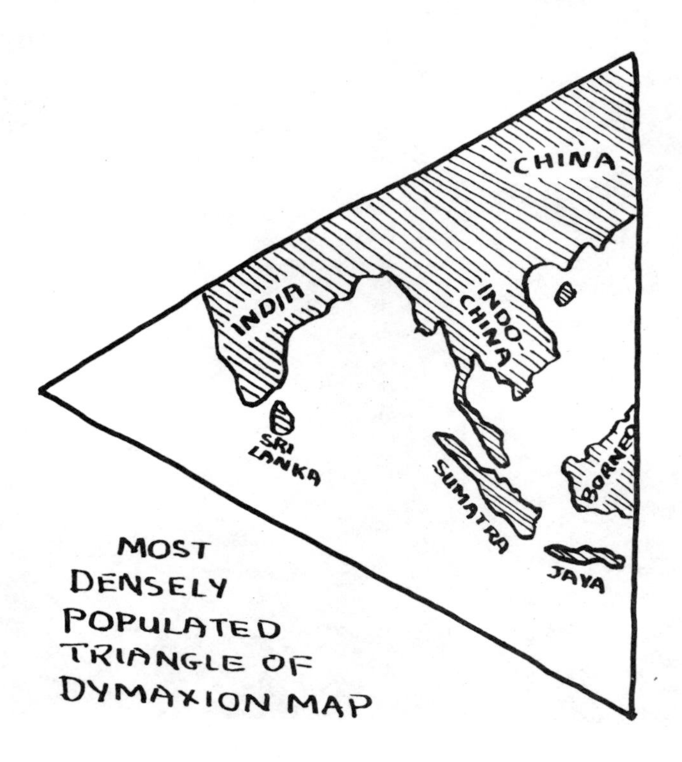

Just look at this one triangle. Here we have 34 percent of the world's humanity in this one triangle—with most of China, Indochina, India, Java, and Sumatra. In North America, which extends over most of *two* triangles, we have only 7 percent of the earth's population. Now two islands, Java and the island nation of Japan, together have 6 percent of the world's humanity—as much as the United States. So you can see that we in the United States are anything but all-important.

Asia contains 54 percent of humanity, Europe and Africa 34 percent, and the Americas 12 percent.

Now for one last concept.

I've already told you that I find the whole educational system is incredibly misinforming, and I can see that your parents unintentionally carry on some of these mistakes. Take this one: "Darling, look at the beautiful *sunset;* isn't it pretty when the sun is *going down.*"

Yet we know that *the sun isn't going down at all.* It isn't setting. The earth is revolving around its own axis to obscure the sun. We all "know" that. But our language has conditioned our senses to the extent that even the great scientists talk about and "see" the sun "set" and "rise." Yet they've known for 500 years that the sun *doesn't* "set" or "rise."

I use two other terms that more accurately describe the reality: *sunclipse* in the evening, and *sunsight* in the morning.

You see, there's no "up" or "down" in Universe. The correct words are "in" and "out."

"Up" and "down" are words from the days when people believed the earth was flat, a great plane spreading out to infinity. In such a world, everything was either perpendicular or parallel to the flat plane, and there were only two possible directions, up and down.

But the real Spaceship Earth is a sphere. And you understand things by tuning *in* or tuning *out,* by converging or diverging.

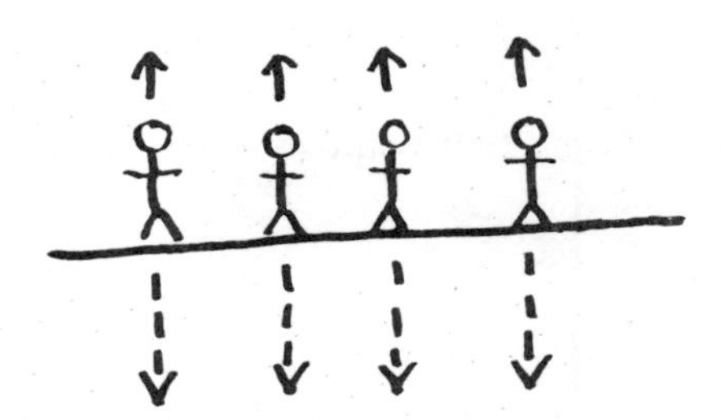

In the mythical flat world, all people standing erect would be exactly perpendicular to the same plane and parallel with each other. Up and down are the only possible directions of movement in relation to an object moving away from or toward the plane.

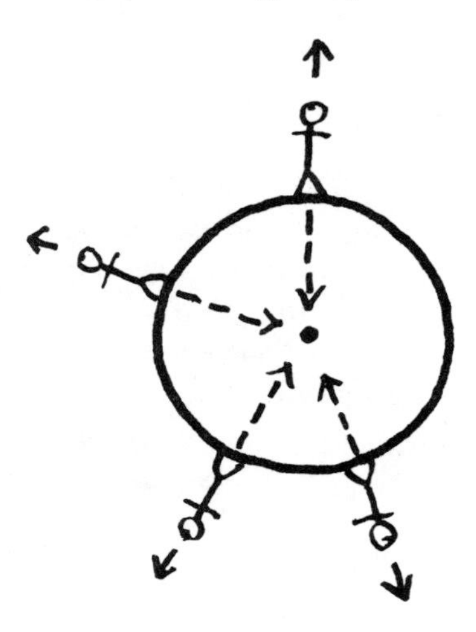

In the real Spaceship Earth, all people standing erect are in a unique relationship to one common center of gravity, from which they can converge (as in digging a mine) or diverge (in an elevator or rocket).

In the real world, all humanity is related to one common center; while in the mythical world, there is only the vast, endless, and imaginary plane.

I think that's enough of my ideas to get you started thinking on your own.

Now I understand that you have some questions for me. I'd love to hear them.

PART FOUR
Dymaxion Dialogue

These are the children's own questions, formulated by them on the basis of three Saturday meetings during which I gave them my own best understanding of Bucky and his ideas. Some of the questions were asked immediately following Bucky's presentation, which you have just read. The others were asked during two subsequent sessions.

BENJAMIN: You seem so much against school; what do you think we ought to have instead of school?

FULLER: We need just to explore and be, and tell ourselves what we're finding. That's what every child starts to do on his own. What parents or teachers are obligated to do is to provide faithful, demonstrable answers.

RACHEL: What were your first thoughts as a thinker? The first thoughts you ever had?

FULLER: I would say that the earliest thoughts I can really remember were about snow. I can remember being terribly excited by snow.

When I was born, we lived seven miles outside Boston, which was really country in those days. You couldn't see any other homes from our house. I remember you could see the glistening of the snow on the trees, and the few little marks made by the feet of birds and squirrels. The very first thing I remember thinking about was the mystery of "How far does this glistening white world go?"

This was a little different than just *looking* at things. It's when you're stimulated to think about what this actually *means.* That white seemed to go off in all directions. It was winter, and there were no leaves, so you could see through the trees. You couldn't see so far in the summer. The white of the snow seemed like a base. You could see farther, it seemed, and you felt you could just go any direction at all on that snow—like you could glide over it.

I was born in July, and this was probably my first Christmas when this happened. I was wondering, "Where does this all go?" This was my first *thinking* thought. Thinking has to do with mind, and with thinking about relationships. It's more than just *reflexing.*

I find, looking back, that it was things that had to do with distances that really made me think. Later on, those things made me wonder as I looked out of an upstairs window, "Where does this all go?" And the same thoughts led me to climb trees so I could see farther, see more.

My memory goes quite far back in life. I can remember my closest friend and I being wheeled side by side in baby carriages.

But what about *you?* How far back can you remember?

RACHEL: I think when I first started remembering things was when I was about a year and a half. I was already walking. I remember pouting about having to sleep with my new baby sister. I wanted a room of my own, and I didn't want to sleep with Shira. I was mad about having all that attention taken from me.

BENJAMIN: If you could rearrange the world in any way you wanted to, how would you change it?

FULLER: I wouldn't change it one iota.

BENJAMIN: Why not?

FULLER: Because I think it's great. You and I don't know enough to change it. We didn't design the human being, and we didn't arrange being on Spaceship Earth.

BENJAMIN: If you could change politics, what would you do?

FULLER: I wouldn't change that either. I don't like politics, and I don't pay them any attention. I think that humanity has fooled itself into thinking it can do more with politics than it really can. By inventing the electric light and making it so people could work at night, Edison did more than anyone could do with politics.

JONATHAN: In another hundred years, will we have the same education system as we do now?

FULLER: From my own experience, I would think that we would not. In our conversation today, I was able to give you in an hour and a half some of the most important fundamentals.

I would say that in the future, we human beings will be able to dial up any information we want from satellites and such. We'll have access to whatever information we need, wherever it is in Universe.

I don't think you'll be going to school; you'll be able to dial up whatever you need right at hand, wherever you are.

BENJAMIN: If we're not going to school, how can we learn things like creative writing? Because computers aren't able to judge creative works.

FULLER: Probably through two-way television. These things can be beautifully taught over television.

RACHEL: How did your schoolmates react to the kinds of questions you asked your teachers? How did they take it when you really *questioned* things?

FULLER: I tended to just get a laugh. I was thought of as a kind of goof. A funny goof.

JONATHAN: You certainly turned out to be a brilliant man.

BENJAMIN: I understand that you believe there's enough food and energy, enough resources, for everyone, and that everyone should have what they need. If that happened, wouldn't there be people who would just be laying about and doing nothing—that is, if things were given out freely?

FULLER: Well, I think there are a lot of people who already do that. There are a lot of rich people who do nothing else. I see nature developing over time, and it probably takes sometimes two or three generations to get enough experience to set things up so a human being can do something. Everybody on the team is not making a touchdown. It takes a whole lot of enormous social activity for any given thing to happen evolutionarily. But I think it will happen.

JONATHAN: You say there's already enough to go around, and that most people feel forced to go to jobs they don't like out of a fear that they need to "earn a living." When do you think man will evolve to a state of mind where he doesn't feel forced to go to a job he hates?

FULLER: There is enough right now for everyone to have a high standard of living. But it has to be organized. In other words, you have to get distribution, and you have to produce certain goods to be distributed. It would take ten years to get everybody actually enjoying this high standard of living, if we all decided that's what we wanted to do.

Today we have the words "earn a living" because we've been assuming there's not enough for everyone. If there's not enough for everyone, if there's not enough to go around, you have to *earn* the right to get in on the life support, the food, energy, shelter, and so forth.

But if we create the distribution, you would never have to "earn a living" again. You would be doing whatever you do because that's what you *want* to do, and not because somebody else says that's what you've *got* to do to earn a living.

The minute you stop equating what you do as part of "earning a living," you'll be doing things because you see something that needs to be done and you feel like doing it; because you like to do it well and want to demonstrate that you can do it better than anyone else. So whatever people do will not be done because of "earning a living."

RACHEL: If all the world were to decide that they were going to live under one government and they didn't know what it was going to be yet and they came to you for advice, what would you tell them?

FULLER: That's a very good question. I'd point out the following two things:

First, we've developed rockets that can put a satellite into fixed orbit around the earth. The satellite can be held in one position relative to the earth, and then be used to beam electronic waves back and forth over the planet. And we can also use these satellites to hold all kinds of information.

Russia and the United States have also learned to use satellites to spy on each other, and on other countries. They have developed things called sensors which can provide incredibly detailed information about things going on aboard the surface of Spaceship Earth. To give you an idea of how sensitive these sensors are, consider that the highest an airplane can go is about

100,000 feet, less than twenty miles. These satellites are more than five times higher than that, and they are so sensitive they can tell the difference between a sheep and a goat. It's amazing that we have sensors so delicate they can make that kind of distinction.

Second, we've also discovered that every human being has an electromagnetic field. I'm very certain that phenomena like what we call telepathy or mind reading are caused by very, very, ultra-ultra-high-frequency electromagnetic waves. That's what accounts for what happens when we're just *sure* we're going to see somebody, and then we turn the corner and there they are. This has happened to so many human beings that I'm sure we all have this ability. It's operating in all of us.

So, we've learned that every human being has an electromagnetic field, although it's very, very high frequency and there's very little energy involved. And when you're happy and pleased, you give off a positive field, and when you're feeling unhappy and displeased, the field is negative.

Now, let's put these two discoveries together, the satellite sensors and the human energy fields. When we do, I would say they might give us a world government that would work like this:

We will *hire*—not elect—people who are skilled at handling mass information, and who have a good voice and appearance for broadcasting. We will have these people presenting propositions [ideas to be voted on] to the whole of humanity.

Now human beings in general—anyone and everyone—will be writing up propositions about what ought to be done about various things. If enough people write in about a particular thing they feel ought to be done, then the management, the hired people, will write up a formal proposition for presentation to everyone. Then on certain hours and minutes which have been well publicized in advance around the world, the television will broadcast the propositions: "We propose to do such and such." And as human beings around the world hear the propositions, they will react positively or negatively—and then our satellite sensors will be able to read their electromagnetic fields and report back to us, "Sixty-seven percent of humanity is for this, eighteen percent against, and fifteen percent is not reacting." So you will have a direct world-around readout instantaneously.

Now we will have set some level of agreement that will have to be achieved before an action can be carried out, say 75 percent. So if 75 percent of the people feel something should be done, you undertake to do that. The function of the world government is to carry out those decisions.

Now maybe the people were wrong, for oftentimes people are wrong. But if they are, you find this out very, very quickly, because things won't be working out. And if things don't work out, people would reverse their positions very quickly, and the world-around satellite sensors would pick up that the majority didn't like what they had done. So other propositions would be put forth to improve the situation.

The whole process works like the steering device for a ship that we call a "servomechanism." That's a piece of equipment that constantly makes corrections to compensate for changing winds and currents to keep a ship on its course.

So if you've decided to do something and you find it's wrong, you do something else.

You see, we walk right foot–left foot, right, left. We don't walk with the same foot forward all the time. All progress is a wave phenomenon. We don't have any straight lines in nature. So, as we make more and more mistakes, we're also learning more and more about what *does* work, and we learn to steer better and better.

This is the kind of government we would be having. Do you follow me?

RACHEL: Right.

BENJAMIN: So you believe there's such a thing as mental telepathy between two people?

FULLER: Ben, I don't use the word "believe" for me. When I say the word "believe," it means accepting an explanation of a physical phenomenon without any experimental evidence to support the explanation.

When I speak of telepathy, I am inclined to accept it by my own direct experience of a number of intuitions, and there have been quite a number of times when I have had telepathic experiences. So my own experience convinces me something is really going on.

But most importantly, there is what happened with our first child. My wife and I were married in the time of World War I. Our first child, Alexandra, was born just at the end of the war, and she caught spinal meningitis and infantile paralysis. Because of these illnesses, she couldn't move around like other children and had to spend her time in a bed being wheeled around. We had two trained nurses, my wife, and myself to look after her.

She couldn't get around like other children, and that meant she couldn't experience things the same way they did. If another child saw something across the room, the child would get up, touch it, turn it around, taste it, and what have you. The normal child could get the *feel* of it directly to build up its inventory of experiences. But our daughter couldn't do that, and so she began getting her information through the people around her who were touching the things themselves.

To our amazement, we discovered she was extremely sensitive to other persons, their feelings, and so forth. My wife and I would be about to communicate—I would be about to say something to my wife—and the words would come out of my daughter's mouth as she lay over there in her bed. My wife and I would both turn, for she wouldn't be using her kind of words, but ours.

So I feel that nature has fail-safe circuits. If one circuit loop stops working, she has an alternate circuit. One of the circuits nature had was for human beings who couldn't get around yet still had to get information. And our child had this circuit, this fail-safe mechanism. I think we all have it.

So when it comes to telepathy, if it's not magic—and I don't call it magic—there has to be some kind of explanation. If it's physical, it's physical. And I know it does happen. So I'm assuming it's an ultra-ultra-ultra-high-frequency electromagnetic wave phenomenon which doesn't work over very great distances (because the higher the wave frequency, the more the possible interference).

BENJAMIN: Do you believe this ability will be more developed as time passes?

FULLER: Right. I think one of these days we'll be able to make direct experiences, to be able to actually measure the energies involved.

BENJAMIN: Do you believe that's for good or bad? Do you think the world will use that for good or bad?

FULLER: Again, I don't use the word "believe" for me. I also don't have any words "good" or "bad." I think if it's part of Universe, it's good. If it isn't part of Universe, it doesn't happen.

There are things that balance other things. If something moves over here, then something else moves over there to compensate for it, to balance it. That's the way I see it. I don't know any human being who has done something society considers obnoxious or undesirable that, had I been born and brought up under the exact same conditions, I would not have done in exactly the same way. So I don't blame the individual about things. It's the conditions. And if we change the conditions that made people act certain ways, then they won't act that way any longer.

JONATHAN: Do you think there's an absolute life form or intelligence—a God?

FULLER: Very much. I'm absolutely convinced of it. Look at the discovery of the law of gravity—the fact that all physical bodies in Universe, all mass, is interattracted to all other mass in a relationship that can only and precisely be expressed in mathematical terms. Both the law and its discovery are purely intellectual, mental phenomena. The law must be eternal, because it's always been there. So what we are really doing is continually discovering these purely intellectual *a priori* (that means here before you, always here) proofs of an eternal intellectual integrity greater than that of humans. So I'm overwhelmed by the evidence, and I would call that God—the Greater Intellectual Integrity we see operating eternally around us and everywhere.

JONATHAN: Do you think there are Martians, or any other intelligent life forms in the universe?

FULLER: I'd say that you and I, designed the way we are, looking the way we do, are uniquely fitted to our own special biosphere, the conditions aboard the surface of Spaceship Earth.

Our bodies are more than 50 percent water. If I hit myself on the arm, hydraulics distribute the impact to other parts of the system. Hydraulics are non-compressible; you can't squeeze water into a smaller space. And that helps us keep our shape. That's the reason I can clap my hands together hard for all these past eighty-five years and not hurt myself. If I were made the same way we make buildings, out of brick and concrete and so forth, I would be just like a china doll and break up if I got hit.

So we're very dependent on liquids, and liquids boil and freeze within very small limits. So we could live only in our particular biosphere where the temperatures are such that we don't boil or freeze. That tells me our kind of design is only for this particular planet.

What intelligent life does is to serve as an information gatherer and local problem solver in Universe. What form that would take to operate on Mars is probably not tune-inable by our particular senses. I wouldn't expect somebody looking like we do on Mars, because he wouldn't need to look like a human being. He would just get the information in other kinds of ways.

JONATHAN: Could you build a city on another planet, then take and cover it with one of your geodesic domes and make it airtight so people could live there?

FULLER: You could do it, but I think it would be very illogical to try. I don't know what we would want to be doing on that other planet. We don't need any more metals; all the metals we need are just coming out of the melting up of yesterday's now-obsolete machinery. We don't need to do any more mining, so I don't see any point in going to the moon or elsewhere for metals.

JONATHAN: Do you think we'll ever travel to other planets? And if we do, what will the spaceships be like?

FULLER: As we learn more about ourselves and the atoms of which we're made, I think we'll be developing ways of scanning—really scanning you in depth, not just on the surface—to identify your own unique individual pattern. And then we could send you by radio; 186,000 miles per second—the speed of radio waves and light—would be much faster than any of the rockets.

RACHEL: What do you think the universe is?

FULLER: My concept of Universe, darling, is the following: In the first place, when I use a word like "Universe," I need to describe what I mean, so that you can understand what I'm saying when I use it. So when I use the word "Universe," I mean the sum total of all the experiences of humanity. That's all—everything—you can think of. The *total* experiences. Everything that everyone who has ever lived has experienced right to this moment. I describe Universe as the aggregate (total) of all humanity's consciously apprehended (knowingly experienced) and communicated experience. And that can mean communicated to themselves or others. There's something about communicating an experience that makes it special, because in order to communicate, we have to describe what it is we are experiencing, to put it in concepts others can understand. So that's how I define "Universe."

Now let's look at "Universe" a little more closely. A long time ago, people thought that everything you could see was happening at the moment you saw it, whether it was a burning candle in the same room or an exploding star trillions and trillions of miles away (of course they didn't realize then just how far away the stars were). As far as what the eye could see, humans believed, everything was happening at exactly the same moment. Everything was *instantaneous,* occurring at the same instant. Everything you were seeing was happening *now.*

But less than a hundred years ago, all of that changed, be-

cause scientists discovered that light has *speed.* Although it moves very quickly—186,000 miles per second—it still takes *time* to get from here to there.

With the discovery of the speed of light came a completely different way of looking at things.

It takes about eight minutes for light to get to us from the sun, our closest star. When you look at the sun, what you're seeing actually happened eight minutes ago. If you look at the North Star, Polaris, it's a light show taking place 450 years ago. If you look at the Big Dipper, you'll see a bright star right at the end of the handle, and that's a light show that's a hundred years old. That's how long it took to get to you. And if you look at the belt in the constellation Orion, you'll see two very bright stars. The light you're seeing from one is 1,100 years old, and the light from the other left its star 1,500 years ago.

So, looking at Universe in view of the fact of the speed of light, Albert Einstein told us that what we are really looking at is an aggregate of *non-simultaneous* events, things taking place at all

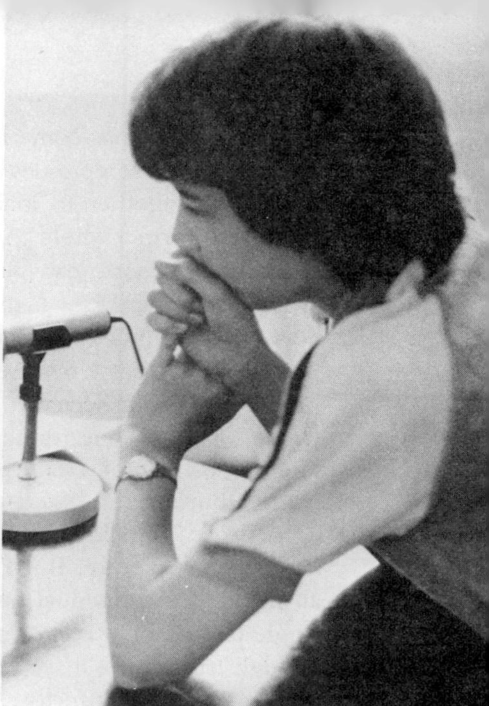

different times. The light just comes to the human eye at the same time, but everything we're seeing is taking place at entirely different times. One of these stars we are looking at may have been happening a million years ago, and the star may now no longer exist.

Einstein told us our Universe is an aggregate of non-simultaneous events. This is exactly the opposite of the "instantaneous" Universe humans once believed in.

Einstein said that each one of these stars is an energy event, and each one involves a tremendous amount of energy; each one has its own endurance, its own span of existence. So each one has an energy transformation going on—stars convert matter into light energy—and each one is taking place at a different time.

The events we see are overlapping one another. One may be coming in as another is going out.

So this Universe Einstein discovered can be described as a *scenario,* like in a motion picture. You have one character come in and he stays for a while and then dies, but before he dies another character has come in, has overlapped the first.

All of these episodes overlap one another, like the threads that come together to make up a rope, overlapping still other threads to form a whole. A rope is a pretty good idea to visualize how a big scenario can be made out of a lot of little scenarios twisted together. Our Universe is an aggregate of overlapping episodes. Beginnings and endings are local to Universe, but they are not Universe. Universe does not begin or end.

Now in our scenario Universe you have individual picture frames, and any one frame doesn't tell you the whole story, just like one picture of a caterpillar doesn't tell you it's going to become a butterfly.

When we talk about Universe, we usually tend to talk about it as a single picture. But I see it as a whole lot of little pictures, all those little individual frames (a "frame" is what you call one single picture on a reel of motion picture film). So after we have enough frames, we'll see our caterpillar going through a cocoon and becoming a butterfly. Then it will take a lot more frames to tell you that a butterfly can fly. To get from the caterpillar to the flying butterfly may take a billion frames.

That's why to get any really important information out of scenario Universe, you have to have large, long increments of episodes. And then you have your episodes and I have mine, and if we tell ours to each other, they overlap and make a new scenario that's a little richer, giving each of us a little better chance to understand what's going on.

Now when you hear that some astronomer has discovered a galaxy that's very far out, you may tend to say to yourself, "I wonder what's outside 'outside'?" But that's a single-frame picture. When you are talking about something that's "outside," that's a single frame. Do you see that?

RACHEL: Yes.

FULLER: So I saw that the difference between yesterday's "instantaneous" Universe and Einstein's non-simultaneous Universe is very great.

You asked me about Universe, and I am telling you that it's

a scenario. The more you and I can talk and communicate to one another, the more we can get to a joint scenario, because we are each adding our information in. That's a beautiful question you asked, darling. I'm beginning to give you some of my best ideas.

BENJAMIN: Do you think the earth will ever become overpopulated?

FULLER: No. I'm absolutely convinced of the wisdom of God.

We've talked a little about the ways things are designed and the way nature needs balances—checks and balances. So sometimes there's quite a lot of mosquitoes, then another year there's a lot of grasshoppers, and so forth. They all interact. They are not "good" or "bad"; they are actually complementary.

Nature doesn't go in a straight line; it's actually more "wavilinear," shaped like waves—like our deliberately non-straight line. Nature has an important function to fulfill and designs something to do it. Surely mosquitoes have some function, and certainly the flies and butterflies all have some function. And when there's a chance of one of her creations not doing well, nature designs another one; she makes more starts.

For instance, on our planet we simply can't take a sunbath to meet all our energy needs. Yet all our energy *does* come from the sun. What happens is that the vegetation—the plants—captures the sun's energy first. By a process called photosynthesis, they impound radiation by using it to create beautiful hydrocarbon molecules which we can use.

Now to be able to capture the radiation for us, the plant has to have roots so that it won't be blown away while exposing all its leaves to the sun. It has to get water through its roots to use in photosynthesis, and to maintain its structure hydraulically. And because the vegetation is rooted and spreads out its leaves to capture sunlight, it has to have some way for its seeds to get planted at some distance away, outside its own shadow, so they can get sunlight and grow. Plants may launch thousands and thousands of seeds, often in a way that they float away on the air or water. Hopefully, some will get out from underneath the shadows of other plants, to places where they will get on all right. But a tree can't deliberately plant its own seeds in a good place; and because the chances are poor of any one seed landing in a favorable place, nature makes many starts—she sends out thousands of seeds.

I say that human beings have very much of a function in Universe. While we are born naked, ignorant, utterly helpless, make enormous mistakes, and need a whole lot of replenishing of us, we're now getting to a point where we have enough information to carry on. We don't need to make so many starts.

Already, birth rates are starting to decline. There have been many predictions that before long there will be ten billion people on earth. We're now about 4.2 billion, and I do not expect to get over five billion. I'm sure we'll never get to ten billion, and that the people who are predicting that large a population are not well informed.

JONATHAN: Human beings have evolved a great deal. Do you think house pets and other domestic animals will evolve to be more than they are now?

FULLER: They tend to evolve more by inbreeding of types, simply because people tend to like, say, a certain kind of cat and begin to inbreed for that type of cat. But cats themselves go out in the night and try to crossbreed back to an average cat.

Evolution is really in the specialist cat, the inbred, hybird cat, which people breed to do certain things that other cats can't do.

If you have two fast-running horses, there's a mathematical probability that by marrying them you'll concentrate the faster genes from both of them so that you might get an offspring that's faster still. This is evolution of a kind, an evolution toward specialization. But what is important about human beings is just the opposite.

If we're really accurate in the way we talk about things, we shouldn't speak of "up" and "down," because those two directions only apply if you live on an absolutely flat plane; while all of us live on a sphere. Instead of up and down, we're either moving out away from the center (diverging) or we are moving in toward the center (converging). This is the way nature operates, by converging and diverging.

Specialization is always diverging from the center, going out and away in one direction from the center. Generalization is coming in toward the center, converging. I realize that we will probably never improve on the design of human beings, because we are already at the center. When people become sick, that's diverging from the center, from normality. Being well is normal, being at the center. Human beings are at the center because they are the least specialized beings in nature. This lack of specialization, coupled with the mind, gives man the ability to confront and solve problems, because he is free to move in any direction from the center; he is not overspecialized, out on some biological limb.

While I think humans have learned a lot, I don't think we've evoluted a lot. Physically we're pretty much the same.

JONATHAN: I heard once on the news that a city was thinking about enclosing itself in one of your geodesic domes. Is this practical?

FULLER: Let me explain it this way. If I take a geometrical object, let's say a cube for the sake of our example, and I want to double its size, then I would double the linear dimension—the length of its edges. If I doubled the dimensions of a cube, I would wind up with a new cube composed of eight cubes the size of the one I started with. [See pages 80–81 for illustrations.]

Where I had one cube, by doubling its dimensions I produce a structure with the volume of eight times the original cube and with faces four times the size of the faces of the original cube. Now 8—the ratio of the volume of a doubled cube to the original—is 2 to the third power, 2 × 2 × 2. And 4—the ratio of the surface of the doubled cube to the original—is 2 to the second power, 2 × 2. So every time I double the size of a cube, I get eight times as much volume but only four times as much surface.

This is also true for the geodesic dome. Every time you double its size, you have eight times as much volume and only four times as much surface. So if I double the size of a dome, I have eight times as many molecules of air inside and only four times the surface. And every time I double the size, I reduce by half the amount of surface through which an interior molecule of atmosphere could gain or lose heat. That means the energy efficiency goes up very rapidly as we increase the size of the domes.

You can see this in the case of an iceberg. An iceberg can melt only as fast as it can get heat through its surface from the outside. Its volume is tremendous, but it melts very slowly because its surface area is so small in relation to the volume. But as it melts, the volume gets smaller at the velocity of the third power, and the surface gets smaller only at the velocity of the second power. In other words, as the surface gets smaller, there is more and more surface exposed per unit of volume, and it melts faster and faster the smaller it gets. The iceberg starts melting only very slowly, speeding up as it gets smaller until you get to the last little ice cube, which melts very fast.

[You can see this for yourself by taking two same-sized ice cubes and two cups filled with warm tap water. Crush one ice cube and leave the other intact. Put the whole ice cube in one cup and the crushed ice in the other at the same time. The crushed ice melts much faster, because far more surface area is exposed per unit of volume, even though the same amount of ice is present in both cups. You will also note that the whole ice cube melts faster and faster the smaller it gets. This is why restaurants often use crushed ice instead of large ice cubes in their drinks. The crushed ice cools the surrounding liquid much faster than an equivalent amount of ice in a large cube.]

Now in buildings, the larger they are the greater their ability to retain energy. In building, the larger the structure the greater the energy efficiency. The efficiency goes up tremendously rapidly as buildings become bigger and bigger.

The difference between a cube and a dome is that the structural stresses place limits very quickly on the size of buildings you can construct based on cubical designs and right-angle corners, while there are no limits on the sizes of dome structures. You could build today a dome that would cover all of Manhattan Island in New York.

I made a study of what would happen if you built a dome over mid-Manhattan. I found that the surface area of all the buildings there was 84 times larger than the surface area of my dome. That means that heat loss in the city is enormous, compared to what it would be with a single dome covering it. It is feasible right now to build any size of dome. Tensegrity makes it possible to put whole communities under cover, and I think people will do it because it's unquestionably much more economical.

RACHEL: How do you think children should be taught the basics of life?

FULLER: Pretty much the way we are doing in this room right now. Try to understand the shape of things. Understand what's around you. Realize that there are mathematical analyses of what's around you. That's an important word, "understanding." When I use that word, it means, "what the mind can do to find relationships." I have been giving you relationships and relationships and relationships.

I would cultivate the child's mind. And that's what the child is already trying to do on its own.

You see a child growing, and soon it comes to the age when it begins to tear things. And it goes to your bookshelves and begins to tear the best books and spares the newspaper lying on the floor. And you say to yourself [Fuller laughs], "Why doesn't it tear the newspaper and spare the books?" But you must remember that the child is constantly experimenting.

You find that children soon learn that if they are lying on a bed and roll over, they can fall off the bed. And they learn when they try to stand that there is an invisible "something" trying to pull them down. They become tremendously aware of gravity, though they don't have a name for it yet. They continue to learn, and they begin to climb up and become a little more sure on their feet. Now when they do this, they want to know what it is they can hold on to when gravity is trying to take over. They want to know what they can hold on to that won't tear.

So what the child is is a research department all by itself, learning in advance what it's going to need to know: what you can hold on to that you can count on.

Now the more you study about children, the more you ask yourself, "What can a child learn and when?" Well, every child is learning about basic principles. If you and I are wise and pay attention to those principles as they are learning, we could have exactly the right things for them to tear, and they would tear them. They would find that this one tears easily, that it takes a little more to tear this other thing; and they would keep on learning until they found something that didn't tear.

You have to say, "This child is *not* a senseless little creature crawling around there; it's a brilliant intelligence, and whatever it's doing is making a whole lot of sense if I really stay with it and see *why* it's doing it."

This is really a very different way of looking at things.

BENJAMIN: You seem to feel there's a difference between the mind and the brain. What is it? What do you use the brain for? The mind?

FULLER: The brain is always coordinating the information of our senses: seeing, hearing, smelling, touching, tasting. It's the only way we know there's something outside ourselves. The brain is the only way we know we're alive. Our brain is our way of saying, "This one smells different from that one." Our brains are dealing in special cases—each sensation we encounter is a special case—and making systems, making packages of what's relevant and irrelevant. Its only concern is the system of the moment. Right now I'm concerned with the system of all-of-us-as-a-group-around-the-table. The brain remembers these special cases.

Now the *mind* from time to time discovers relationships between the special cases that are not in a special case considered just by itself. For example, the discovery of the existence of the solar system, as opposed to looking at just one planet at a time.

Another example of mind is Newton's discovery of the law of gravity, the law of mass interattraction of celestial bodies. This is an existing relationship that does not manifest itself in any single body considered just by itself.

So the mind deals in relationships.

Science is the use of mind to discover *generalized laws.* A generalized law cannot have any exceptions; and because it cannot have any exceptions, a generalized law is inherently eternal.

So mind is dealing in the eternal, and the brain deals in the temporal, the special cases that begin and end. There is a very big difference.

Take the case of Copernicus, a Polish astronomer who lived in northern Italy in the 1500s.* In those days, the power people wanted a universe with earth at the center. But there was no stopping the power of Copernicus' mind. He tracked the position of the planets over twenty-one days, measured to the minute and second. These two sets of facts—the starting and ending positions and the length of time—were like a hinge, a lever. They enabled him to prove that the earth and all other planets moved in elliptical orbits around the sun. His mind's discovery had such a shattering effect on the ways people had been accustomed to seeing things that we still refer to it as the "Copernican Revolution."

Thinking—as opposed to simply reflexing—is done by mind, not brain; brain only coordinates the senses, the special-case circumstances, and gives us our sense of being alive.

Mind is like a director inside a TV studio control room, brain, where all the information from the senses is displayed. Mind compares and coordinates.

* Nicolaus Copernicus, 1473–1543, a Polish-born physician, theologian, and astronomer best known for his *De Revolutionibus Orbium Coelestium,* which is regarded as the foundation of modern astronomy.

You know, it's really amazing. We've never really been outside ourselves. It's as though each of us lives inside an omnidirectional television tube!

JONATHAN: What kinds of recreational activities will we have in the future?

FULLER: I think there will be enormous amounts of new things, because people will be doing more and more what I do—thinking, sitting at desks perhaps too often. So they'll want to go out and get some exercise to keep their body working right. I see more and more of this.

Just in my lifetime, we have gone from where skiing and tennis were sports pursued only by millionaires to now, where suddenly everybody's in on them. So there will be enormous amounts of recreation in the future.

JONATHAN: What kinds?

FULLER: I don't know what the forms will be. I'm confident that where today we have hang gliding, we are going to have new forms of personal transportation tomorrow—especially jet harness packs with two small ramjets. Jet stilts, you might say. With these jets and the right clothing, you'll just go to the window and dial wherever it is you want to go and you'll be off.

That will enable us to do a lot of observing. We'll find it very exciting to look at the land below us as we fly through the air like birds.

BENJAMIN: What do you think the most important culture has been, either present or past?

FULLER: You are inclined to ask me which is most important, but I see things the way Rachel wants to see them—where everything counts.

I don't say *this* is more important than *that*. I see anything and everything that's happening as all built in and everything interrelated. Everything. I can't ever look at things the way specialists do, as separate and not related.

BENJAMIN: Do you have a favorite person? Someone you think best reflects our future?

FULLER: I don't have any favorites. I can see, however, which would be the next person who could help me with whatever it is I need to know next.

I find that when I play tennis, I always play better when I play with a person who's better than me than when I play with someone who's not as good as me. I have often learned to win over a person who's better than I am just because I really play so much better. And I have found I tended to lose games to somebody I ought to be able to win over very easily, because I simply wasn't stimulated to do better.

So instead of having a one-and-only favorite, I can see that I really ought to think about who is next in my series of evolution of stimulations. And then that person becomes my hero for the moment. Then I graduate on to a new hero.

RACHEL: How do you feel after you've completed a project?

FULLER: Your questions are awfully good. Really, they're tremendous questions.

When I finish a project, I always feel tremendously let down. As I am finishing, I'm terribly excited because I've got it done. My projects usually take a whole lot of work, like a book. My last book manuscript was a thousand pages, and I finished just the day before yesterday.

But the minute it's done, where I had been very busy for so long, suddenly I'm not very busy. I really feel terribly let down.

I think it's like a mother on the night after her daughter's been married. Everything has been building up to this wedding, and it's all been terribly exciting. Then, suddenly, the bride's gone and all that's left are a whole lot of dishes to pick up and clean up. So that's the way I feel, darling. I can't wait to get started on something else again.

BENJAMIN: I know you decided to get by with as little sleep as possible, and that you've slept only two hours a day for most of your life. How has this affected you, and would you recommend others do the same?

FULLER: I would never advise anyone to do anything I've done. I'm glad to tell people what I've done. If they find something useful in it for them, fine. But if they want to do it, it must be their own spontaneous act. They may find something that seems worthwhile to them, but I never tell them to do anything.

Now you asked me about not sleeping. This began as an experiment; I was trying to find something out. I had an enormous amount of work to do, and we had no money; although I saw that we always managed to have enough to eat. I decided I had to get the most out of my time.

I saw that cats and dogs go to sleep for a while, then they're ready to spring into action. They didn't seem to take one long sleep, but a lot of little ones.

I said to myself, "I've learned in cross-country running that I have second and third winds; but when I exhaust this third wind, it's energy that took a long time to store away, and that it's sort of remote and hard to get at—like it's 'way over there.' So I should learn to live only on my main tanks, like cats and dogs, and never have to use that remote energy."

Animals can lie down, sleep briefly, then spring into action. When they lie down, they're replenishing their main tanks, because the main tanks have much larger openings and are easier to fill. So I said, "Supposing I live on *my* main tanks?"

Because I had already decided to do what I wanted to do, then I reasoned that should I find myself tiring or my interest flagging, then maybe at that moment I needed to be doing something else: making some tea, dancing, reading something, getting some exercise. And if none of these worked, maybe I needed some rest, and so I'd lie down.

I didn't know how it was going to turn out, but looking back I can see that it worked out that I would lie down and sleep solidly for a half-hour four times a day—amounting to a total of two hours' sleep a day. And I've been getting on great ever since. I find that I've never gotten down to my second or third tanks. And I found out what I was trying to find out—that I could get by on my main tanks with a whole lot less sleep.

But these are *my* conclusions, and they might actually be erroneous. They're what I found out, and I can't recommend it to anyone else.

RACHEL: What kinds of problems do you think will be confronting people in the future?

FULLER: The biggest problem is education. Up to the time I was born, the reality people shared was one where everything we were doing could be seen, heard, smelled, touched, or tasted. You could experience it directly through the physical senses. If you were digging a ditch, you could see it and feel it. When people wrote, it was still about things you could see and smell and hear.

But when I was three years old, the electron was discovered. And from then on, we began to get into an invisible reality. Microscopes and telescopes became much more important. We were beginning to see parts of the universe we couldn't see with the naked eye.

And now today, 99.9 percent of everything that is going to affect our future is being conducted in the realms of reality non-directly contactable by the human senses.

So I think one of the biggest problems, darling, is that we're dealing in a reality which is non-visible and non-touchable by humanity. And if the newspaper can't take a photograph of it, they can't talk about it.

And then everybody is so specialized. That's another of the big problems. Every child, like yourself, is interested in *everything* and wants to understand *all* the interrelationships. But the education system and the moneymaking people—the power system—were afraid of the bright people and made them all specialists, because they didn't want to lose their help. They made them all specialists and said, "You have to pay attention to this little bit over here, but never mind the *whole.*" So the power people kept track of the big things and made all the bright ones specialists to look into the little things, the "special areas."

So the big problem of tomorrow is getting everyone back into comprehensive thinking and understanding, getting all humanity on the right kind of mathematical coordinate system. Those are the kinds of things I've been giving you today. That's why I talked about them right at the very outset.

The problem is getting humanity to really educating itself properly. Once humanity has the right information, it's going to make some very good decisions, particularly through the electronic media.

BENJAMIN: Do you believe we're going to have another war?

FULLER: Again, I don't use the word "believe." With the information I have, I'm inclined to feel that we're having a war all the time. The people are fighting in Afghanistan right now, and this is being carried out by Russia and the United States through its puppets. We've been having a terrible war going on continuously—Vietnam, Cambodia, Nicaragua, El Salvador—I'm sorry to say. I hope we'll have less and less, but if we ever get into an atomic war, then it's all over for everybody.

BENJAMIN: If there is a war between Russia and the United States, will it be carried out by humans or by computers?

FULLER: The Russians have a complete upper hand as far as numbers of soldiers, trained divisions, and they have a stronger navy. Now America has more atomic bombs, but if they go for the bombs, then it's all over for everybody.

No sane person is going to start a nuclear war. But there are a lot of angry hotheads with access to the "button," so it's absolutely touch and go as to whether or not we're going to make it.

Russia's way of doing things is *not* to use the atomic bomb, and Russia now has the upper hand and is beginning to rule the world. The United States had the upper hand until about ten years ago, but now the Russians have it. I think the Russians are most eager to prove their system, their kind of socialist government, can produce a more attractive way of living than the American model. They would like to use their productivity for people instead of weapons, so I think they will insist on disarmament. They'll keep the upper hand, but they themselves will be disarming as fast as they can, with the other people disarming ahead of them. They're eager to prove that socialism can work, so I think we'll get away from a war.

RACHEL: When you were a child and came upon discouragements, were there any children—any friends—who urged you on.

FULLER: No, I really had no urging whatsoever. There was one friend who I felt was incredibly bright—he got the highest marks at Harvard and we were extraordinarily great friends—but even he never urged me on to teach myself anything. And in my earlier days, too, I never really had any friends who tried to urge me on to learn anything. It was all really done on my own.

BENJAMIN: Is there some special way of looking at problems that will help you solve them, help you know what the outcome will be?

FULLER: First of all, I look at problems as for everybody, not just for me. That's number one. And I say that the answer, if there is an answer, is in universe. I also try to think of the significance of it, the significance of the solution and what value it will be to society. How does the problem fit into the scheme of things? Is it an important problem or a non-important one? I give priorities to problems and ask myself, "Is this a problem that I want to tackle today, or can I put it off to tomorrow and work on something more important today?"

If it's one I'm going to tackle today, if it's important to everyone, then I begin to define the problem. The first and most important thing in solving a problem is to state what the problem is, very clearly, concisely, and incisively. State what the problem *really is.* I can't stress how important this is: a problem stated is a problem solved. So many people misstate the problem to fool themselves. But if a problem is properly stated, if its relatedness is clearly defined, the answer will become apparent.

RACHEL: How did your first collaborators react to discouragements?

FULLER: Oh, I've really been surprised. Back in the late 1920s, when I was designing the Dymaxion House, I had a young architectural draftsman working with me who was really terribly excited by me. He said, "Can I come and work with you?" So he came around and worked with me for a while. But then one day he said, "We don't have any income. I'm going to have to leave you. I have a wife and a child, and I'm going to have to get a job."

Well, I had a wife and child too. But I was saying that I don't have to "earn a living" if I'm doing what evolution wants done, what God needs done. And if I'm doing that, then I'm sure a means of keeping me going will occur by itself. I'm doing an experiment to see whether I'm doing what's needed for humans to be a success. We're supposed to be here for a purpose, and that purpose seems to be to make humanity a success. And if I'm helping to do that, then it seems like Universe will keep me going. The hydrogen atom doesn't have to earn a living to act like a hydrogen atom. It's only human beings who seem to be having to "earn a living."

So I said, "If I'm really doing what God wants done, the universe should care for me just as the clover is cared for when the bee comes along and pollenizes it."

But my early collaborators could not see that, and they would go off, get a job, and forget about me.

BENJAMIN: How would you suggest solving international problems without violence?

FULLER: I always try to solve problems by some artifact, some tool or invention that makes what people are doing obsolete, so that it makes this particular kind of problem no longer relevant.

My answer would be to develop a world energy grid, an electric energy grid where everybody is on the same grid. All of a sudden there would be no problems any more, no international troubles. Our new economic basis wouldn't be gold or dollars; it would be kilowatt hours.

RACHEL: How do you think people will look back on your ideas and inspirations twenty years from now?

FULLER: I hope they will find them useful, darling. That's all I ask.

Because I've really been trying to find out how the universe is working, and how to help people to see things, I have an assumption that if we survive this great crisis we're going through right now—and atomic war is a great threat—then people will be pleased that I did what I did.

When I first started fifty-two years ago, I tried to depersonalize what I did. I tried to operate anonymously. But by doing that I attracted even more attention to myself, because people were trying to find out who was doing those things. So I let myself be visible. But I don't do anything for me. I don't do anything because I want to be important or known. Everything is done because something needs to be done for humanity.

As far as I'm concerned, I didn't invent my mind. I didn't invent me. And I'm absolutely overwhelmed by God. I have absolute faith in the Greater Intellect operating. I'm just trying to be a medium for that Intellect. I don't care if after twenty years people discover God and not me. In fact, that would really be something.

You see, darling, we are here in Universe for our minds, not our muscle. Just look at the eruption of Mount Saint Helens. There was more physical power expressed there in one second than has been expressed by all the human muscles throughout all humanity's existence. But man can invent a camera and capture the whole thing.

Mind discovers relationships. That's why we are here in Universe. To discover relationships. To solve problems. And the answers are already here if we can just get ourselves out of the way.

I hope that I've discovered a few relationships that will help humanity.

BENJAMIN: Do you think there is a world energy problem?

FULLER: There is no energy problem whatsoever. We just have a problem of ignorance and fear.

BENJAMIN: How will our country change in the next century?

FULLER: Physically, probably not at all. It will be the same continent in a hundred years.

Let me first say again that I don't know if we're going to get into the twenty-first century. We'll have to pass through a very great crisis.

It used to be an East/West world, with ships being the dominant form of travel. Now it's a North/South world, and we travel by jets. If we make it into the twenty-first century, we'll be

a planet of *world* people. We will be integrating all our people very rapidly.

I've been around the world forty-seven times, and that will be just average for a human being in tomorrow's world. You will all be world people. You may live in Alaska part of the time, then in India, then—anywhere. You'll be *world* people, and that will be the biggest change.

BENJAMIN: Of all events of recent years, say the last twenty years, which one has changed our lives the most?

FULLER: Robert Burns, the Scot poet, once said, "Oh wad some power the giftie gie us/To see oursels as others see us!" That is, how precious a gift it would be to see ourselves as we are seen by others.

And for that reason, I think getting to the moon was one of the most important events of recent history, to be able to be off our planet to see ourselves.

This has changed the total concern of humanity, helping everybody to become comprehensivists, to be interested in the whole universe and not just your local South San Diego, or whatever.

BENJAMIN: What stories did you like most when you were young?

FULLER: I was most excited by Robin Hood. I was intrigued by the idea of an outlaw trying to solve things for people where he didn't think justice had been done. I loved the daring, and the fact that his life was a terrific fight. But then I began to realize that Robin Hood was using weapons—bows and arrows and swords—and that I could do a better job with trigonometry. I changed the Robin Hood game.

I also like the King Arthur stories, the *Faerie Queene*, Shakespeare, Dickens, the old English tales. I think reading aloud in the family is very important. Libraries are very important.

JONATHAN: By disagreeing with your teachers, did you get bad grades?

FULLER: Oh, no. I knew what they wanted me to answer, and I gave them their answers. But I said to myself, "I'm going to have to keep at things I've noticed which seem true to me, and discover for myself if there is any significance to them."

I found that teachers were not interested in *my* explaining things to *them*. It really upset them, and I decided that I was not trying to upset people.

BENJAMIN: What books would you recommend we read to learn more about your ideas?

FULLER: My experience with you makes me answer that question completely differently than I would have ten years ago. There is nothing that I talk about, however intricate it appears to me, that you have not been able to talk about with me. Therefore, I would suggest that you read some of my books. The kinds of things I've been talking about with you are the kinds of things I've been writing about.

I think you would enjoy *Operating Manual for Spaceship Earth, Critical Path,* and my other books. I think you would enjoy *Synergetics* very much.

BENJAMIN: Are there any other questions you'd have liked us to ask?

FULLER: You couldn't have asked any better questions if you were ten or twenty years older. I'm sure no professor could have asked better questions. I had no idea you would ask such good questions. Really, it was very exciting to me that you asked them. You children were asking me the kinds of questions I'd ask myself. They were very good.

RACHEL: I was thinking one last thing. Let's say we start to make our own colonies on other planets and that we start to make our cities there. Would it be possible for there to be another advanced civilization that was already there, but from another dimension, so that we could both be on the same planet but each not know the other was there?

FULLER: In other words, they just happened to be invisible to us, to our spectrum of vision?

RACHEL: —and we happened to be invisible to their spectrum.

FULLER: That really is possible. And it may be true throughout Universe, and not just on one planet. You see, I emphasize in every way I can that whatever life is, it is not physical.

JONATHAN: I agree.

FULLER: I'm now eighty-seven, and so far I have eaten, drunk, and breathed in eight hundred tons of food, water, and air —which became temporarily my hair and got cut off, became my skin and got rubbed off, and so on. Everything in me changes every seven years. I am not yesterday's breakfast. That's very clear to me, although my breakfast does temporarily become my hair.

We used to have the words "animate" and "inanimate." In playing the game of twenty questions, where you had to find out what I was thinking about by asking me no more than twenty questions, two of the questions were always, "Is it animate?" or "Is it inanimate?"

When we said animate, we'd think of warm and soft; inanimate was hard and cold.

Now we used to say that biology was the study of animate things, of living things. After a while, biologists got onto genetics, to studying genes and chromosomes (the microscopic chemical chains that determine the inherited traits of organisms).

The biologists who were looking into genetics began studying fruit flies, because they regenerate very rapidly and thus gave the biologists a chance to find out how patterns of development were inherited. Then they discovered that tobacco mosaic virus regenerated even more rapidly than the fruit fly, and this brought them into virology, the study of viruses. And inside the viruses they discovered that the chemicals were arranged like crystals.

Now crystals had always been considered as inanimate, lifeless. But here the physical characteristics of living things, of you and me, are being determined by these crystal-like chemicals called DNA and RNA. And these chemicals are composed of molecules, and they in turn are formed of atoms.

Now at this level you have all the scientists converging. The biologists, the physicists, the mathematicians are all there studying the same thing. But they've become so excited about what they're discovering that they haven't been very good philosophers. They haven't seen the significance of what they're dealing with.

We're seeing more and more things that are inanimate, and just what is animate is getting less clear. We know that life has to be in there somewhere, because we started with biology, with living things. But when people die, all that's left is chemicals. People have been weighed as they were dying, and scientists have found that the weight doesn't change. So whatever life is, we know it doesn't weigh anything.

I'm assuming that nobody ever did die, and that others whom we can't see may be trying to talk to us right now. I know that there's really great wisdom that's available to us that can affect us a great deal, and that's why I try to listen to my thoughts, thinking that others may be trying to talk to us. That may be the way we get our fresh thoughts.

TO EVERYONE: It's been good to be with you. I feel as though we were old friends. I enjoy being with you so very much.

POSTSCRIPT

Richard Buckminster Fuller died July 1, 1983, eleven days before what would have been both his eighty-eighth birthday and the sixty-sixth anniversary of his marriage to his beloved Anne. He died of a heart attack suffered at Anne's bedside in a Los Angeles hospital. Anne died two days later.

I spoke to Benjamin Mack the morning after Fuller's death.

"When I heard last night that he had died, I remembered he was talking about life and death and reincarnation at our last meeting," Ben reminded me. "And you know, I'm sure that Bucky is still going right now, doing whatever he's supposed to be doing."

PART FIVE
Through the Eyes of a Child

After the first session, I asked the children to write a paper, any length and in any style, describing their day with Bucky. The results were delightful.

MY DAY WITH BUCKMINSTER FULLER

By Jonathan Nesmith

I can simply start by saying that Mr. Fuller is quite a remarkable person. I'm just sorry that *I* didn't make a decision like his (a decision not to believe everyone, in order to find not necessarily more *correct* answers, but more trustworthy ones—for he came to the conclusions himself).

When we—Mr. Brenneman (the writer), Mr. Myrow (the sound man), Ben, Rachel (two other kids approximately my age), and I—were first introduced to Mr. Fuller, it was quite ordinary, simple, and relaxed. Later I found, not only was he not *ordinary* or simple, he was a very brilliant person with an extraordinary concept of things.

As he stood in the doorway, I quickly looked at him. He was wearing a pair of glasses and had very short hair that was a whitish color. Before we were ready to record the conversation, I must admit I was a little nervous. To be honest, about a week before I met Mr. Fuller, Mr. Brenneman asked me to write a few questions for Buckminster Fuller. He was sort of conducting an experiment. Mr. Brenneman wanted Mr. Fuller to express his ideas to younger children. At the time, I didn't even know who Buckminster Fuller was. I felt this experience was no major event. Not only was I wrong, but when we got out of the car to go into his house, I shook like a leaf.

In actuality, when we arrived Mr. Fuller wasn't home, so we walked around the block. However, I still shook. Mr. Fuller looked like an elderly man, quite elderly in fact. I soon found out he was eighty-five years old. This, of course, did not affect his agility—not mentally. We all entered his house, which seemed normal except for a few things such as little decorations. We were all seated around a round table. Mr. Fuller excused himself and left the room for a second. In the meantime, Mr. Brenneman assembled and situated all his camera equipment, and Mr. Myrow hooked up his rather large tape recorder and placed a microphone at the edge of the table. Mr. Fuller then entered the room carrying two bags. He laid them down on the floor. The materials that occupied the bags consisted of foot-long sticks and three-inch red rubber tubes. With these you would be quite capable of building geometric figures.

I began thinking he was going to build a geodesic dome (which he invented) right before our very eyes! Of course that was silly of me. He wouldn't do a thing like that. He already had made one. In front of us children were papers covered with questions we were prepared to ask him. Apparently he didn't even know what they were, for he said "What are those?" We all spoke up at once saying basically that they were questions about the future we'd like to ask him. All the questions were good questions, even though mine were written slightly sloppily. But Rachel's were "humdingers," so to speak. Mr. Fuller asked if we wouldn't ask them at the moment, so I folded my paper and slipped it in my pocket.

He then began telling us how his teachers taught him some incorrect things. In order to illustrate that point, he assembled a square from the sticks and tubing. He announced that a square doesn't exist and illustrated this by holding up the figure he had made. It wouldn't hold its shape. The sides kept moving and the angles kept changing. It would be almost impossible to find the surface area, because you could be capable of making this square have a million different surface areas. This fact was amazing, and ever since that day, I couldn't get over it. Mr. Fuller took even more sticks from the table and assembled a cube. Sure enough, it collapsed over on its side. But then I began to wonder if it was possible to make this stable. I looked at the cube and found you would be able to stabilize it. In order to do this, you would just place a cross on each surface of the cube, so I asked Mr. Fuller if a cross would do this. Mr. Fuller told me that even one stick placed diagonally would do it. I must say Mr. Fuller trusts thoroughly in triangles.

A little while before this subject came up, Mr. Fuller had told us about the tetrahedron and had assembled one from his sticks and rubber tubing. He removed one stick from the base of the tetrahedron and asked me to reassemble another similar figure. When I had completed it, he requested that I place the tetrahedron upside down inside the other one. I scrutinized the figure and realized that essentially this figure was a cube. We discussed this for a while and soon we all felt it was time for an intermission.

During this break we all ate lunch except Mr. Fuller, who went off to another room while we ate. Most of my conversation consisted of very few words. When I opened my mouth, about the only thing that came out was "Isn't that neat how there's no such thing as a square? Isn't it?!!!" Other than that I would say something like "Please pass the salt." As we tidied up after ourselves, Mr. Fuller entered the room.

By then everyone was ready to go back to it, so we all sat down at the table. Mr. Fuller said he was prepared to answer our questions, so we all took out our papers. Rachel began, and the questions went around the table until I had to ask one. Before I met Mr. Fuller, Mr. Brenneman had told me about his distrust in some of the things his teachers taught. So as a result, I asked the question "Will we have the same educational system that we have now a hundred or so years from now?" He didn't think so. He said that by that time there would probably be computers that we would go over to and type out a question and it would answer. As the questions went around again, I thought of this. It would be amazing to fit so many ideas and answers into a computer no bigger than, let's say, a typewriter.

Once again the questions came around to me and I spoke up, asking, "Will we have the same recreational activities that we do now?" He replied with "A long time ago, you had to be extremely wealthy to play things like tennis or golf. Soon a jet pack will be developed that anyone can connect to their back. Then they can shoot around the sky." I thought this was incredible. I wish they would hurry up and invent it! I thought.

We finally finished up with questions. Then Mr. Fuller gave each of us a geodesic globe that you assembled by yourself, and I had him sign the back. After Mr. Fuller went to his room, Mrs. Fuller entered and said, "We have a lovely pool in the back. Would you like to go swimming?" I thought it would be fun, so I accepted, but then I realized I had no swimming trunks. "That's no problem," she said. "You can wear a pair of Bucky's shorts." So I tried them on. They were slightly big, so in order to make them fit I pinned them tighter. So I swam a bit and had a wonderful time. All in all, the day was fabulous and I hope to meet him again and really I can simply end by saying Mr. Fuller is quite a remarkable person.

BUCKY

By Ben Mack

I remember Bucky as a sweet, lovable man. However on my way there I had doubts. Before we even got in the car to go to his house I felt unsure and shaky. When we arrived early I was quite nervous but wanted to cover it up. As we walked over to a bakery I began to feel more and more sure of myself. By the time he opened his door I forgot I was nervous.

Bucky was very outgoing and we got to know him (and his wife) quite fast. When we first sat down he couldn't get our names straight because he's hard of hearing, but we quickly jumped that obstacle.

He was quick to start a conversation, so we got straight to work. As we started to talk we noticed how much there was to say. He talked about math and school first. I got a little bewildered with the information that he gave us because there was so much of it. The more he talked about school, the more he talked about his childhood. I think when he talks to people it is a form of telling his history.

When we broke for lunch there was a sense of unclarity, because of all that was happening and a nervousness of whether it was going all right. We had a big lunch and then resumed our discussion. This time there were open questions.

Questions after questions were asked on all subjects, including politics. I enjoyed this time best, but I don't think Bucky liked it as much. Every type of question was asked, but he liked the ones where he could expand on to several topics, like "What were your first thoughts as a thinker?"

Bucky also liked it when people questioned what he said to answer our questions. He greatly encouraged questions and that was how he became what he was. I agree with him, but people can find you irritating if you ask too many questions.

I felt very satisfied when it was all finished, but now that I know what he's like I have more questions I would like to ask him.

His house was located near the beach. It was not out of the ordinary or modernized in any visible way (like I had expected). Instead it was a ordinary house and decorated normally. The neighborhood he lived in was quiet and not very busy. In other words, if you just look in his living room, dining room, and kitchen you wouldn't guess it was his house. Despite his house Bucky is a very special and unordinary person. I had a very special time and I would just like to say so.

AN AFTERTHOUGHT

The day we spent with Buckminster Fuller was very important to me. He denied just about everything I had learned in geometry from my school. Most importantly, he gave me a new way of looking at the world. He feels that what is happening on "Spaceship Planet Earth" is good and that everything in the Universe is good (and if it's not in the Universe, it doesn't exist). It's just that sometimes things could be a little bit better.

I agreed with a lot of what he said; however, sometimes I didn't. In my opinion, people take the wrong approach of always saying yes to things they don't understand. I didn't understand everything he talked about, so I don't know if I agree with him on all of his theories. So don't let anyone fool you by saying $E = mc^3$. It might be wrong.

Our day with Bucky added a very special addition to my education. I hope that those who read the book that will tell about our experience will gain as much from it as I did. I am more aware of the many possibilities in the universe now that Bucky has shown us new portholes to look through. It is a world full of promises.

A DAY OF LEARNING AND EXPLORING WITH BUCKY

By Rachel Myrow

Saturday, June 14, 1980, Dick Brenneman took Ben Mack, Jonathan Nesmith, and me to Bucky's house. As we were a little early, everybody, including Dick and Daddy [Fred], took a walk. We got back. Bucky greeted us warmly and showed us to some seats and a table where we could discuss and explore. When we were settled and Dick and Daddy had set up their photographic and tape equipment the meeting began.

First Bucky began discussing the main principles of his explorations. He talked about the fact that when the teacher explained things that she only showed one side of the subject and never encountered the other side of the blackboard. He explained that some people didn't notice that fact. Then he took his pen and paper and began showing us some of the main laws and facts of nature and how one may find those facts and how they worked. Like Mother Nature works on triangles and hexagons. And the law of doing more with less. When he had gotten about an hour and a half of information into our heads we changed into testing that information by building models that Bucky provided. We examined the models and studied the fact that the square would flop and the triangle would remain stable.

For lunch Ben's mother had dished up a variety of foods from yogurt to chicken. I think that lunch was well appreciated.

After lunch Bucky began answering questions. Several times Ben asked questions using the word believe. It was then that we learned that when you believe a fact it means that you *think* you know the fact but you haven't *tested* the fact. One of my questions was "How did your collaborators react toward major misfortunes?" The answer: "They didn't understand that when you're working for the Lord's purposes, you have everything you need." Jonathan asked many questions about the future, and Ben asked questions about all sorts of things, and everyone came out of the meeting with satisfying answers. Bucky and Anne suggested swimming but we had brought no trunks so we ended the day by Dick and me swimming in Bucky's trunks.

It had been a fun day of Learning and Exploring with Bucky.

PART SIX
Summary and Reflections

The children spent three days with Bucky, over the space of two years. The results of those sessions have been distilled in the pages you have just read. I guess that leaves me with the final word.

The most obvious difference between the first and later sessions was in the attitudes of the children in the moments before they rang the bell at Bucky's house. Prior to the first session, the children were nervous, self-conscious, feeling burdened by a sweaty-palm-instilling sense of being responsible for "making good." In the moments before the second session, the children were eager, playful. Indeed, they formed a three-member chorus line as they trooped through Fuller's door, each sporting a piece of impromptu "geodesic" headgear.

Bucky had become a friend.

Before the first session, each of the youngsters had spoken of "Mr. Fuller." Before the next two sessions, it was strictly "Bucky."

If there was a single moment when their common love of learning shattered the barriers between Bucky and the children, it was when he proved to their complete satisfaction that "there is no such thing as a square." It was this one point that the children discussed on the drive home. They had become common conspirators with Bucky, aware that the emperor—geometry as they had learned it in school—had no clothes. From the moment of that awareness, Bucky was assured of his audience.

Bucky was hard of hearing and would strain to catch each word. The children often had to repeat their questions, gradually gaining the confidence to speak up as the sessions progressed. Between question and answer, Fuller often paused, sometimes closing his eyes as he waited for his thoughts to form.

Some of the questions were quickly answered; he had been asked them hundreds of times before in a life punctuated by encounters with curious and intelligent listeners. Some of the questions were more provocative, demanding longer reflection and often raising a smile at the unexpected delight of hearing something new.

Bucky imparted a tremendous amount of information, particularly in the first part of the initial session. "Don't worry if they don't seem to get it all now," he advised me at the end of the day. "They'll find it for themselves when the time is right."

I agree.

I know that the children received two precious gifts during their sessions with Bucky. First was Fuller's acceptance of each as an equal, as a person just as gifted, unique, and worthy as himself. (To Fuller, each child is a born genius, and innate brilliance is diminished only through the imparting of erroneous information.) The second gift was his deep sense of trust in their own intuitions. "Listen to what your inner voice is really telling you," he seemed to be saying. "What you *feel* may really be true, despite all the world's noisy arguments to the contrary."

The sharp black-and-white of the printed page can never capture all the character of the encounter between Buckminster Fuller and his newfound friends. From the transcript may spring the impression that Bucky treated Benjamin and Jonathan a bit harshly, especially concerning the word "belief." But Fuller's voice was never raised; indeed, he seemed to be offering a midstream semantic guidance correction (as he might say), aimed at an erroneous thought rather than an erroneous person. Nonetheless, Fuller was distinctly taken with Rachel's questions, and with the fact that she had told him she was interested "in everything"—a comprehensivist, Fuller described her in subsequent speaking engagements and interviews. (Ben had mentioned an interest in magic, and Jonathan in electronics.) Bucky attributed

Rachel's broad interests to the fact that she was born after humanity had reached the moon, and the two boys were born before. It was typical of Fuller to find broad meanings in specific events; however, I think he may have overgeneralized here.

I believe Bucky has helped to inspire the children to persist, to test what the world says about each experience against their own intuitions. By his own example and his faith in them, Bucky imparted to each child the sense that he or she was as capable as anyone of offering significant gifts to humanity. And he offered a healthy antidote to the self-obsession and jingoism that seem to have infected American culture in recent years. Help others, he urged, and you'll find your own needs taken care of.

It's my own conviction that the life of each child was deeply and beneficially touched by these unique encounters and that more "education" took place in those few hours than in years of conventional schooling. Ben Mack in particular was enthralled with Bucky's synergetic geometry, so much so that he began large-scale model making and has perfected his own connecting system for octahedrons, and invented a new connector for assembling geodesic domes. His mother has told me the encounter has imparted new purpose and focus to his life.

Bucky inspired me, too.

Long before our first meetings, I had been following his life. I was fascinated by his inventions and discoveries, and even more by his *ideas.* But most of all, I was inspired by his vision of humanity and of the immense potential of each individual.

As a journalist, I concurred wholeheartedly with his desire for precise use of the language. Words, I knew, could obscure as well as illuminate; and it was the responsibility of the journalist to describe events concisely and clearly. I was particularly intrigued with Fuller's insistence on banishing the terms "up" and "down" from everyday language. These two erroneous terms, he declared, conditioned the very way we perceive ourselves and our relationship both to earth and to universe.

After the first session, I set out to prove for myself if it might be possible to experience myself differently if I looked at earth, sun, moon, and stars as Bucky urges—as a human looking *out* from the surface of spherical Spaceship Earth.

Repeatedly, on different occasions, as I gazed heavenward at the celestial orbs, I struggled to perceive myself as looking "out" instead of "up."

It worked.

Suddenly, on a drive in the Mojave Desert, there came a moment as sun and horizon began to merge, when I really *was* looking *out* from the surface of Spaceship Earth. I found myself *feeling* for the first time a passenger on a great sphere hurtling through the cosmos. Venus was just coming into view, and the nearly full moon was at the eastern horizon. Sun, moon, and planet described the great arc of the ecliptic. At that instant I *knew* the location of poles and Equator. I felt a sense of *place,* of proper relation, that I had never known before.

My awareness of the world, the whole universe, was revolutionized, transfigured, in an instant. For the first time, my *felt experience* of reality was coinciding with what my intellect had long known to be true. It was an initiation, a rite of passage. I felt for the first time a citizen of the cosmos. I was no longer tied to a language-conditioned flat earth.

And there was a sense of communion with all humanity, with all living things, in the knowledge that we were all related through one common center, earth's center of gravity, all passengers on an infinitely precious star-faring vessel.

I know others who have shared the same experience. It is joyous, in that something old is suddenly seen in a new light. It is awesome, because it affords a glimpse at a reality far grander than we have been conditioned to perceive. And it is sobering, because it reveals how deeply conditioned (mesmerized, if you will) we can all be by habitual patterns of language and thought.

"I'm only a low-average human being," Bucky professed. "The only difference between me and other people is that I decided to devote my life to discovering just how much a low-average human can do."

His conviction—and the challenge it implies to each of us—was (and is) sincere.

Time will prove or disprove the validity of Bucky's theories and formulations. But his faith in the individual was transcendent and infectious. I saw him share it with three children, and each caught the spark of his vision.

BIBLIOGRAPHY

All books, unless otherwise noted, are available from the Buckminster Fuller Institute, 1743 S. La Cienega Blvd., Los Angeles, Calif. 90035. Write them for a complete list. They also offer Dymaxion maps, charts of Fuller's synergetic geometry, and much more.

BY R. BUCKMINSTER FULLER

4D Timelock, Fuller's first book, written in 1928 soon after his fateful decision. Reprinted by the Lama Foundation, Albuquerque, N.M., 1972. Paper, $5.

Nine Chains to the Moon, published in 1938 and containing Fuller's early thoughts on housing, ecology, and industry. Doubleday, Garden City, N.Y.; and Southern Illinois University Press, Carbondale, Ill. Hardback, $7; paper, $2.95.

Education Automation, containing Fuller's vision of the educational system of tomorrow, 1962. Doubleday and Southern Illinois University Press. Paper, $1.95.

Ideas and Integrities, consisting of autobiographical musings and thoughts about the future. Collier Books, New York City, 1962. Paper, $3.95.

No More Secondhand God, poems and essays, Southern Illinois University Press, 1969. Paper, $2.95.

Operating Manual for Spaceship Earth, consisting of the most easily read statement of Fuller's vision about the earth and man, and what they can become. Southern Illinois University Press, 1969. Hardback, $4.25; Touchstone Books, New York City, $3.95 paper.

Utopia or Oblivion contains a series of lectures and addresses by Fuller, including his unique perspective on human historical development. Penguin Press, London, 1970, hardback, $7.95; Overlook Press, New York City, hardback, $11.95.

50 Years of Design Science Revolution and the World Game, historical documentation (articles, clippings) plus commentary by Fuller, 1969. Paper, $5.

Earth, Inc., more thoughts along the lines of Spaceship Earth, 1973. Peter Smith Pub., Magnolia, MA, paper $7.50.

Synergetics: Explorations in the Geometry of Thinking. This massive and difficult but rewarding book will challenge almost every assumption you have about how things work. It's the distillation of Fuller's revolutionary geometry, his magnum opus. Macmillan, New York City, 1975. Hardbook, $35; paper, $12.95.

Synergetics 2: Further Explorations in the Geometry of Thinking, a complementary volume to *Synergetics,* containing color drawings and an index (essential) to both volumes. Macmillan, 1979. Hardback $27.50.

And It Came to Pass—Not to Stay, containing statements by Fuller of social and political concepts. Macmillan, 1976. Hardback, $7.95.

R. Buckminster Fuller on Education, a collection of previously published essays on education, including the full text of *Education Automation.* University of Massachusetts Press, Amherst, 1979. Paper, $6.95.

Critical Path contains a common (relatively) language statement by Fuller of his view of humanity, its history, and what can happen if we all pull together. St. Martin's Press, New York City, 1981. Hardback, $15.95; paperback, $9.95.

The Dymaxion World of Buckminster Fuller, by Fuller and Robert Marks, contains a profusely illustrated inventory of Fuller's inventions and dreams. Anchor Press, 1973. Paper, $5.95.

Buckminster Fuller: An Autobiographical Monologue/Scenario, by Robert Snyder. Fuller's son-in-law, a filmmaker, records Fuller's words and images in a comprehensive biography of Fuller and his thought. St. Martin's Press, 1980. Hardback, $15.95.

Tetrascroll, Fuller's retelling of the Goldilocks tale, cast as a parable in synergetics. St. Martin's, 1982, $15.95 hardback, $7.95 trade paperback.

Grunch of Giants, a sequel to *Critical Path.* Fuller's urgent summation of the forces behind the current global economic crisis—and a scenario for its transformation with the inadvertent assistance of multi-national corporations. St. Martin's, 1983, $8.95 hardback, $4.95 trade paperback.

Inventions, a fully illustrated catalog of Fuller's 27 patented inventions and the last book he submitted to his publisher. St. Martin's, 1983, $40.

BOOKS ABOUT R. BUCKMINSTER FULLER

Cosmic Fishing: An Account of Writing Synergetics with Buckminster Fuller, by E. J. Applewhite. A former CIA official gives his account of helping Bucky write his most formidable book. A wonderful look at Bucky's human side. Macmillan, 1977. Hardback, $7.95.

Pilot for Spaceship Earth, by Athena V. Lord. A biography written for children. Macmillan, 1978. Hardback, $7.95.

The Mind's Eye of Buckminster Fuller, by Donald W. Robertson. Fuller's patent attorney describes the joys and travails of Fuller's genius. St. Martin's Press, 1983.

OUT-OF-PRINT WORKS

Untitled Epic Poem on the History of Industrialization.

Fuller's poetry came about because many of his sentences
proved so long that they "read" best
broken into segments like this

to make comprehension easier
and to give the eye relief.

The untitle speaks for itself. Simon & Schuster, New York City, 1962.

I Seem to Be a Verb, with Jerome Agel and Quentin Fiore. Little text, lots of pictures. A 1970 McLuhanesque experiment. Bantam Books, New York City. Paper.

Intuition contains poems and thoughts of a philosophical nature. The title also reflects the name of Fuller's beloved sailboat. Doubleday, 1973.

INDEX AND GLOSSARY

This is both an index and a glossary. The definitions provided are consistent with Fuller's perceptions. For more on the geometrical terms, see Fuller's *Synergetics* and *Synergetics 2,* which provide comprehensive explanations of his synergetic geometry. For more on Fuller's thinking about humanity, see *Critical Path* and *Grunch of Giants.*

When a word within a definition appears in italics, refer to that word's separate entry in this index/glossary for a definition. Italicized page numbers signify illustrations.

A

B

C

D

E

N

O

P

T

U

V

W